D1712168

Reconstitutions of Transporters, Receptors, and Pathological States

Efraim Racker

Division of Biological Science
Section of Biochemistry, Molecular and Cell Biology
Cornell University
Ithaca, New York

1985

ACADEMIC PRESS, INC.
Harcourt Brace Jovanovich, Publishers
Orlando   San Diego   New York   Austin
London   Montreal   Sydney   Tokyo   Toronto

ACADEMIC PRESS, INC.
Orlando, Florida 32887

United Kingdom Edition published by
ACADEMIC PRESS INC. (LONDON) LTD.
24–28 Oval Road, London NW1 7DX

Library of Congress Cataloging in Publication Data

Racker, Efraim, Date
    Reconstitutions of transporters, receptors,
and pathological states.

    Bibliography: p.
    Includes index.
    1. Biological transport.   2. Cell receptors.
3. Diagnosis, Cytologic.   I. Title.
QH509.R33   1985   574.87'5   85-11280
ISBN 0–12–574664–4 (alk. paper)
ISBN 0–12–574665–2 (pbk. : alk. paper)

PRINTED IN THE UNITED STATES OF AMERICA

85 86 87 88   9 8 7 6 5 4 3 2 1

Contents

Preface

Lesson: Those who today do not work with DNA should have their DNA examined. Those who clone for the sake of cloning are biochemical clowns.

The first lecture in an earlier book I wrote was entitled "Troubles are good for you." I have been urged by a young scientist to retract this concept. Because of serious problems that beset my laboratory a few years ago I have seriously considered this proposition. But once again I became convinced that it is up to us to decide how to face trouble. Once again we are enjoying what is happening in the laboratory. According to a Chinese proverb "When the dust passes thou will see whether thou ridest a horse or an ass." When the dust settled over our troubles, I discovered that I was riding a mule and I could not expect it to gallop. I shall describe in Lecture 11 what emerged from our troubles and what progress we have made studying transforming growth factors.

Having thus reconfirmed that "troubles are good for you," I would like to transmit another unpopular viewpoint to our students. I am fully aware of the extraordinary possibilities of the recombinant DNA approach and I understand the attraction it has for young students, young professors, and even aging professors. During the DNA crisis almost ten years ago I expressed my views about the importance of this field and about the irrational fears of an epidemic caused by an artificially created Andromeda strain. But I did not recognize at the time the danger of another threatening epidemic—an epidemic of profit or expected profit that lures some of our best students and faculty away from university departments to companies that are primarily interested in the sale of products. The directors of these companies are very bright, and they want to keep the recombinant DNA research challenging and exciting. They emphasize its fundamental aspects and give some investigators great freedom—for a while. This, combined with the glitter of gold, is hard to resist. The gold rush of recombinant DNA that has resulted in the mushrooming of companies that cannot all survive will lead to disillusions and disappointments. I want to convince our graduate students in this book that there are many exciting adventures still ahead of us in other fields that need to progress with and without the help of recombinant DNA.

This book—like my previous ones—is based on outdated lectures that I have tried to bring up to date as I revised them. The onslaught of new pertinent publications finally became too much for both me and my secretary and I made an arbitrary full stop. Once again I want to extend my apologies to those colleagues whose work I should have quoted but did not. I hope they will be sympathetic to this problem, drowning like the rest of us in a flood of publications that are difficult to keep up with.

I have neglected important aspects of reconstitutions involving planar bilayers and patch-clamping and have referred to excellent reviews that have been published. I have not discussed the elegant work on the reconstitution of ribosomes, on protein complexes involved in muscular contraction, on the assembly of enzymes and viruses, and many others.

I acknowledge with gratitude the financial support I have received from the National Science Foundation, The American Cancer Society, and particularly the National Cancer Institute. I wish to include in this rather impersonal acknowledgment to institutions a note of thanks to the administrators who have gone out of their way to help me in recent years of financial restrictions.

Once again I want to thank all of my students and collaborators who helped me overcome my troubles during the past few years: M. Abdel-Ghany, P. Boerner, E. Blair, S. Braun, R. Feldman, B. Hilton, A. Kandrach, E. Lane, J. Lettieri, S. Nakamura, J. Navarro, M. Newman, W. Raymond, R. Resnick, C. Riegler, K. Sherrill, D. Stone, D. Westcott, J. Willard, L-T. Wu, X. Xie, and Y. Yanagita. I want to express my gratitude to Mike Kandrach, who for almost thirty years has kept my students and postdoctoral fellows and their instruments in good repair. I am indebted to Judy Caveney, my secretary, who was willing to help again after the trauma of the previous book, and to Melissa Stucky, who has aided her. I acknowledge valuable comments and suggestions, particularly by Dr. Gottfried Schatz, Dr. Nathan Nelson, Dr. Piotr Zimniak, Dr.

Dennis Stone, and Dr. Gregory Parries, during the preparation of the manuscript.

Last, but not least, the inevitable thanks to my wife Franziska and our daughter Ann, her husband John, and their children, who often kept me away from the journals, thereby exerting the necessary restrictions on the size of this book.

Efraim Racker

Lessons

Abbreviations

AchR	Acetylcholine receptor
AIB	α-Aminoisobutyric acid
A23187	An ionophore for divalent cations and protons
AMP-PNP	5'-Adenylyl-(β,γ-imido)diphosphate
Bio-Beads	Polystyrene particles used for removal of detergents
c-*src*, c-*myc*, c-*ras*, etc.	Cellular oncogenes (protooncogenes)
cAMP	Cyclic AMP
cAMPdPK	cAMP-dependent protein kinase
CF_1	Chloroplast coupling factor 1, chloroplast ATPase
CL	Cardiolipin
ConA	Concanavalin A
CoQ	Coenzyme Q, ubiquinone
$C_{12}E_8$	Dodecyl octaoxyethyleneglycol monoether
DAO	n-Dodecyl-N,N-Dimethylamine oxide
DCCD	N,N'-Dicyclohexylcarbodiimide
DIDS	4,4'-Diisothiocyano-2,2'-disulfonic stilbene
DOPC	Dioleoyl PC
DOPE	Dioleoyl PE
DTNB	5,5'-Dithiobis-(2-nitrobenzoic acid)
DTT	Dithiothreitol
EAT	Ehrlich ascites tumor
EGF	Epidermal growth factor
ER	Endoplasmic reticulum
Extracti-Gel D	Particles used for removal of detergents
F_1	Coupling factor 1, mitochondrial ATPase
F_2 (Factor B)	Coupling factor 2
F_6	Coupling factor 6
$Gpp(NH)_p$	GMP-PNP, 5'-guanyl-(β,γ-imido)diphosphate
5-HT	5-Hydroxytryptamine, serotonin
kDa	Kilodaltons
KNRK	NRK cells transformed with Kirsten virus
LDL	Low-density lipoprotein

MDCK	Madin–Darby canine kidney
MeAIB	α-(Methylamino)isobutyric acid
NBD-Cl	7-Chloro-4-nitrobenz-2-oxa-1,3-diazole
NBMPR	Nitrobenzylthioinosine (6-[4-nitrobenzylthiol]-9β-ribo-furanosyl purine)
NEM	N-Ethylmaleimide
NP-40	Nonidet P-40
NRK	Normal rat kidney cell line
Octyl POE	Mixture of octyl-polyoxethylene (3–12 ethylene oxide units)
PDGF	Platelet-derived growth factor
PK	Protein kinase
PPdPK	Polypeptide-dependent protein kinase
PMS	N-Methyl phenazonium methosulfate
PTS	Phosphoenolpyruvate : sugar phosphotransferase system
PE	Phosphatidylethanolamine
PC	Phosphatidylcholine
PS	Phosphatidylserine
PI	Phosphatidylinositol
RGC	Transporters containing recognition protein (R), GTP protein (G), and catalyst (C)
RCR	Respiratory control ratio
SDS–PAGE	Sodium dodecyl sulfate–polyacrylamide gel electrophoresis
SITS	4-Acetamido-4′-isothiocyano-2,2′-disulfonic stilbene
SR	Sarcoplasmic reticulum
System A	Na^+-dependent amino acid transporter A (alanine, methionine, etc.)
System ACS	Na^+-dependent amino acid transporter ACS (alanine cysteine, serine, etc.)
System L	Na^+-dependent amino acid transporter L (leucine, methionine, etc.)
TDX	Tetradotoxin
TGF	Transforming growth factor
v-*src*, v-*myc*, v-*ras*, etc.	Viral oncogenes
1799	Bis(hexafluoroacetonyl)acetone
3T3	Mouse embryo fibroblast cell line

Lecture 1

Resolution and Reconstitution of Soluble Pathways and Membrane Complexes: Overview of Principles and Strategies

What a good thing Adam had—
when he said a thing
he knew nobody had said it before.

Mark Twain

I. Reconstitution of Soluble Pathways

Resolution and reconstitution is a classical approach of biochemists to the mysteries of intact cells. In 1927 Otto Meyerhof added a fractionated yeast extract to a crude muscle extract. Neither preparation alone fermented glucose to lactate; together they did (Fig. 1-1). The muscle extract contained the enzymes that fermented glycogen; the yeast fraction contained an enzyme which "activated glucose." This is how hexokinase was discovered, and this is how hexokinase was first assayed in a reconstituted system. It illustrates the two purposes of reconstitution: a method of assay and an approach to the analysis of the whole from its parts. During the subsequent years, one glycolytic enzyme after another was separated and studied in isolation. Finally, highly purified glycolytic enzymes and the required cofactors were put together and shown to catalyze steady-state glycolysis, provided an ATPase was added (Gatt and Racker, 1959). The insight into the role of ATPase in glycolysis was again a contribution by Otto Meyerhof, who in 1945, as a German

1

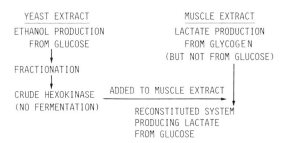

Fig. 1-1 The first reconstitution experiment.

refugee in Philadelphia, performed some of his last experiments (Meyerhof, 1945). He added a partially purified potato enzyme that hydrolyzed ATP to an extract of dried yeast that fermented glucose poorly. If the appropriate amount of ATPase was added, steady-state fermentation was observed, but if either too little or too much ATPase was added, the utilization of glucose ceased after a burst of activity. Meyerhof concluded that the ATPase must be "in step" with the hexokinase. For each molecule of glucose that was phos-phorylated by hexokinase, two molecules of ATP must be hydro-lyzed to deliver the appropriate amount of ADP and P_i required for the oxidation of glyceraldehyde 3-phosphate (Fig. 1-2).

In the same year it was observed (Racker and Krimsky, 1945) that a homogenate of mouse brain did not glycolyze in the presence of Na^+ because of an excess of ATP hydrolysis. Glycolytic activity was restored by slow infusion of ATP or by addition of phospho-creatine (Table 1-1). Thus, a heterologous ATP-generating system was introduced into a biological pathway to rectify the imbalance caused by excessive ATPase activity.

The first reconstitution of a complete pathway was based on the work of Horecker and our own group on the enzymes that partici-pate in the reductive pentose phosphate cycle (Racker, 1955). Gly-colytic enzymes isolated from yeast and muscle were combined with enzymes of the pentose phosphate cycle isolated from yeast and spinach leaves. In the presence of ATP and $NADPH_2$ the mixture fixed CO_2 and converted it to hexose.

Lesson 1:   How to reconstitute unphysiologically

Reconstitution experiments performed by combining en-zymes from muscle, yeast, and spinach, can hardly be called

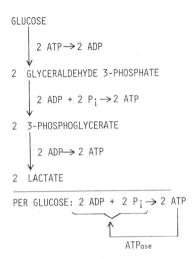

Fig. 1-2 "ATPase" is a glycolytic enzyme.

physiological. We should designate Otto Meyerhof as the father of unphysiological reconstitutions. What is a daydream to the biochemist may be a nightmare to the physiologist. But the only way we can find out whether the whole is the sum of the parts is by putting the pieces together again and learning how they work. It is the task of the physiologist to help the biochemist by pointing out what is missing. It was the physiologist Otto Meyerhof who knew what sources of enzymes to use to achieve the best reconstitution of a pathway.

TABLE 1-1 **Stimulation of Glycolysis of Brain Homogenates in the Presence of Na⁺ by an ATP-Regenerating System**

Additions to the cofactor mixture	Glycolysis (μmol per hour)	
	Δ Glucose	Δ Lactate
None	−11.3	+21.1
Phosphocreatine (5.5 mM)	−15.3	+24.9
NaCl (40 mM)	− 0.5	+ 2.4
NaCl + phosphocreatine	−12.4	+24.1

The next task was to see whether physiological control mechanisms can be observed in reconstituted systems. We added mitochondria to a reconstituted glycolytic pathway and observed phenomena resembling those first described by Pasteur and Crabtree (Fig. 1-3). We found (Wu and Racker, 1959) that glycolysis is inhibited when mitochondrial oxidative phosphorylation competes for ADP and P_i (the Pasteur effect) and that respiration is inhibited when glycolysis dominates and deprives the respiratory chain of ADP and P_i (the Crabtree effect). These experiments convinced us of the importance of P_i and ADP as rate-limiting factors in bioenergetics (Racker, 1965, 1976), a mechanism of regulation that was first recognized by M. Johnson (1941) and, independently, by F. Lynen (1941). Both in mitochondrial respiration and glycolysis the generation of ADP and P_i can be a rate-limiting step. In mammalian cells the ATP-generating machineries are present in excess, geared to energy utilization. What an ingenious and simple way of food economy! Energy is generated only as it is needed.

The experiments on the Pasteur effect were extended (Uyeda and Racker, 1965). The role of the allosteric inhibition of phosphofructokinase by ATP and hexokinase by glucose 6-phosphate was demonstrated in reconstituted systems of glycolysis. These important control mechanisms regulating the utilization of sugar do not change the Meyerhof stoichiometry. For each mole of lactate that is formed, one mole of ATP is generated and must be hydrolyzed to maintain steady-state glycolysis. To understand the driving force that propels glycolysis, the ATP-hydrolysing processes, referred to broadly as ATPases, must be identified.

It will be shown in Lecture 11 that the major contribution to ATP hydrolysis in tumor cells takes place in membranes. Thus, glycolysis, which is dependent on the availability of ADP and P_i, is a mem-

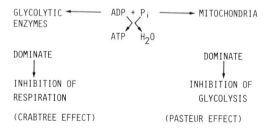

Fig. 1-3 Reconstitution of the Crabtree and Pasteur effects.

brane-dependent process. A membrane-free cell extract does not glycolyze unless an ATPase is added (Racker *et al.*, 1984). It is a reflection on the individuality of cells that the relative contributions to ATP utilization by the plasma membrane and organelle membranes vary considerably and are related to the specific function of the cell. We are just beginning to gain an insight into the balances of budgetary expenditures in bioenergetics.

Many biochemical pathways have been reconstituted by combinations of isolated enzymes participating in the formation and degradation of purines, pyrimidines, fatty acids, and amino acids, as well as in the biosynthesis of macromolecules. A great deal has been learned from studies of reconstituted multienzyme systems.

Questions that can be answered by performing experiments with reconstituted soluble multienzyme pathways

1. Are the known components sufficient to catalyze the overall pathway?

2. What are the kinetic interactions between functional neighbors?

3. How do manipulations of individual enzyme concentrations change the multiple rate-limiting steps and the susceptibility of the pathway to inhibitors?

4. Which enzymes are controlled by allosteric regulators and how do they affect the operation of the overall pathway at different enzyme concentrations?

The pathways mentioned above are catalyzed by water-soluble enzyme systems that do not require a compartment for function. The next task was to explore membrane-bound pathways. In this lecture I shall present a broad outline of the various methods that have been used in the resolution and reconstitution of membrane components and recapitulate what we have learned from these experiments.

II. Resolution and Reconstitution of Membrane Complexes

A. Why Do We Do It?

Why do we want to destroy the marvelous architecture of a natural membrane? The answer is basically the same as we give to justify the breaking of cells to study soluble multienzyme systems. We need to identify the working parts of the machinery involved in physiological functions.

A young student may ask today whether reconstitution of membranes is an important area of research. Are there still problems left to be solved in membranology; is it a challenging and an exciting field? The answers are yes, yes, and yes. Until now most of the systems that have been attacked were those concerned with basic mechanisms of cellular transport and even those need much more work. Moreover, we have begun to move from broader to more specific physiological problems. How much do we know about the transport systems in various anatomical sections of the kidney or the intestine? How much do we know about the functions and regulation of individual brain cells? The field of export and import of macromolecules is still in the stage of infancy. We know little about the role of the cytoskeleton in the plasticity of the cell membrane, and we need to know more about intracellular mobility of macromolecules and organelles. How do macromolecules move in and out of organelles? How are they excreted? What are the chances of progress in these areas of research?

New methods of resolution and reconstitution have been developed. They need to be refined and expanded. We are still lacking in methods for directional reconstitutions, eliminating the formation of mixed populations of right side out and inside out vesicles. In the future we must develop methods for reconstitution of compartments of the size of cells so that we can insert organelles and the cytoskeleton. We need to refine and extend the emerging method of reconstituting organelles that I shall discuss in the last lecture.

We also have to approach problems of differentiation and reconstitution of specific cellular function (e.g., proton excretion in the kidney). Great advances have been made in methods for growing differentiated cells in defined media in cell cultures that will facilitate approaches to specific organ problems. Collecting large quantities of cells from culture flasks is still laborious and expensive, but we can expect continuous improvements in this technical area. It will be possible to grow many more primary and secondary cultures of highly differentiated cells in large quantities using specific inhibitors (e.g., monoclonal antibodies to eliminate undesirable competitors such as fibroblasts).

The challenges facing us in the exploration of specific hormone and drug action, internalization and secretion of proteins, cytoskeleton function, cell division, differentiation, and in the analysis of pathological membrane processes, seem countless. I believe that these problems cannot be solved without some help from reconstitution.

B. What Problems Shall We Choose, What Tissues Shall We Select, What Assay Shall We Use?

A problem must be ready for attack. I shall mainly discuss approaches to problems in the field of transport that are susceptible to attack by reconstitution. There are many others; the rules are similar.

Lesson 2: *How to choose a transport problem*

Remember that phospholipid bilayers are impermeable (a) to most charged and many uncharged compounds; therefore, anything that is charged yet readily crosses natural membranes must get across via a transport system and (b) to macromolecules; any protein that enters or exits a cell must involve a transport system. Most membranes have pores with selectivity controls.

Thus, if you are interested in movements of ions or proteins across the membrane, join the transport union and the reconstitution club. Membership is free. New members are welcome but viewed with suspicion.

Pick a problem that is not too fashionable, for example, pick a problem involving the intestinal tract. Few biochemists have a taste for it. There are probably more compounds that are metabolized and get absorbed in the intestine than are listed in the Merck Index. If you happen to find a toxic compound that is not absorbed, it could be a valuable drug against worms.

There are a few guiding lessons which we should impress upon a beginning research scientist. Let us suppose there is a young M.D. who is interested in kidney functions. She or he may wish to explore how Na^+ is reabsorbed in the convoluted tubules or how protons are released in the distal tubules of the kidney. Perhaps he or she has already performed some interesting physiological experiments on acid secretion in tubules isolated from rabbit kidney. Will it be feasible and prudent to use the same material for biochemical reconstitution studies? Will it be possible to make progress with a few milligrams of starting material isolated from 30 rabbits?

> *Lesson 3: Choice of an abundant starting material for the preparation of membranes*
>
> If you are rich (you have a large NIH grant): use cows.
> If you are not very rich (you have an NSF grant): use a few bushels of spinach.
> If you have very little money: use human placenta.
> Don't use (unless you live on the West Coast) *Torpedo californica.*
> Hesitate to use (unless you own a company that clones silver dollars) tissue culture cells.

Once the choice of starting material has been made and suitable supplies of membranes have been accumulated, you will have to decide on an assay. Many transport functions are activated by specific ligands or are sensitive to specific inhibitors. In the past, such ligands and inhibitors have been successfully used not only in binding assays during purification of the protein but also in the development of specific affinity columns. These are most useful methods, but we must be aware of serious pitfalls associated with an assay depending on ligand binding. I shall describe these later. In the final analysis, a functional assay needs to be developed and this is sometimes easier said than done. Yet, without it, we may prepare a pure protein which is dead and which may have been severely altered in the physicochemical properties during purification.

> *Lesson 4: The assay*
>
> Don't use a binding assay if you can help it. Spend many months developing a functional assay. If you don't have a functional assay for the crude protein, use a binding assay until you can develop a functional assay.
>
> *Remember that:* A dead acetylcholine receptor was isolated when α-bungarotoxin binding was used as an assay. A dead lactose transporter was isolated when a protein labelled with radioactive N-ethylmaleimide was purified. A "functional" assay with proteoliposomes also can be misleading unless it is representative of the physiological process. For example, a Ca^{2+}/Na^{+} exchange was observed after reconstitution of a Tri-

ton extract of excitable mambranes. However, the vesicles did not exhibit inhibition by external Na$^+$, characteristic of native vesicles. In contrast, proteoliposomes prepared by reconstitution of a cholate extract did show this regulatory phenomenon.

A specific inhibitor or activator of function is a valuable aid in staying on course during purification of a membranous protein.

C. Isolation of Enriched Vesicle Preparations

Before approaching the destructive process of membrane dissolution with detergents, it pays to (a) search first for a readily available source that is rich in the transporter or receptor under study, and (b) attempt enrichment of the protein by fractionation of the membrane and isolation of enriched membrane fragments. This was the approach taken by Changeux and his collaborators in tackling the problem of the acetylcholine receptor. Starting with the physiological information of the abundance of acetylcholine receptor in the electric organs of fishes, they developed an isolation procedure that yielded vesicles greatly enriched in acetylcholine receptor (Sobel *et al.*, 1977). Later it was shown (Neubig *et al.*, 1979) that a prominent protein that appeared in SDS-gel electrophoresis as a 43,000-dalton band could be selectively removed by extracting the membrane fragments at pH 11.5 without damage to the receptor. This kind of approach will become increasingly appropriate when dealing with mixed membrane fragments from organs (e.g., kidney) containing a mixture of cell populations. The use of affinity columns made with monoclonal antibodies or with specific ligands help in the separation of wanted from unwanted membrane fragments. For the isolation of plasma membrane vesicles with different orientations, ConA or wheat germ columns have been used which adsorb vesicles that are right side out. More specific examples for these approaches will be described in subsequent lectures.

Lesson 5. Preparation of membranes and their solubilization

Remember the individuality of membranes. An enzyme in one membrane may have properties different from the same enzyme in another membrane. Choose a membrane in which "your protein" is present in huge amounts. Search for the best source. Remove as many "impurities" as possible *before* solubilization (use mild detergents, alkali, or acids). Impurities are defined as everything you are not interested in.

For solubilization use a detergent which will solubilize at least 70% of your protein and leave most "impurities" behind.

Note that the above is easier said than done. "Your protein" is probably a minor component, sensitive to acid, alkali, and most detergents. Moreover, no one has succeeded in solubilizing it. (That is why you are trying it.)

It should be realized that the fact that one can purify specific membrane activities without detergents is inconsistent with the original model of the fluid mosaic membranes. The observation that one can isolate without the use of detergents membrane fragments that are enriched in bacteriorhodopsin or Na^+,K^+-ATPase is a clear indication for the presence of strongly associated functional complexes that do not readily move laterally in the phospholipid bilayer.

D. Resolution and Reconstitution from Without

I would like to illustrate this approach with several examples, some ancient and some more recent. The approach of resolution from "without" is based on the fact that many membrane-associated functions are catalyzed by complex structures that are made up of multiple and different subunits. Some of these are located within the hydrophobic domain of the phospholipid bilayer; others are outside the bilayer, linked to the membrane by forces of various strengths and characteristics.

When thylakoid membranes of chloroplasts were isolated, the ability to reduce NADP was lost. It was restored by a protein in the soluble fraction which was loosely associated with the membrane (San Pietro and Lang, 1958). This was the beginning that provided an assay and led to the isolation of ferredoxin.

The energy transformer F_1 of the mitochondrial ATPase complex is so loosely bound to the membrane that shaking submitochondrial particles with glass beads (Pullman et al., 1960) removes F_1 from the membrane (Fig. 1-4). In chloroplasts, exposure to a low ionic strength in the presence of 1 mM EDTA (Avron, 1963) allows for the dissociation of CF_1, the chloroplast ATPase. Other coupling factors (attachment factors) are more firmly bound. Some of them can be pried loose by sonic oscillations. Treatment of the particles with silicotungstate (Fig. 1-4) was used to remove several coupling factors (F_2, OSCP, and F_6), while the membrane structure was basically

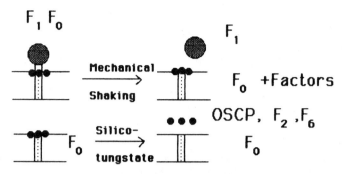

Fig. 1-4 Resolution of coupling factors.

left intact (Racker *et al.,* 1969). To separate hydrophobic components, more drastic methods involving detergents were required that
will be discussed later.

Another and more recent example is in the field of protein secretion (Blobel and Dobberstein, 1975; Meyer and Dobberstein, 1980;
Meyer *et al.,* 1982). Secretory proteins pass from the cytosol across
the membrane of the endoplasmic reticulum (ER) into the lumen.
Often, but not always, this passage is accompanied by the proteolytic removal of a "signal peptide." Initiation of protein synthesis
occurs in the cytoplasm on free ribosomes. After about 70 peptide
bonds have been formed, the signal sequence has usually emerged
from the large ribosomal subunit. A cytoplasmic ribonucleoprotein
complex called "signal recognition particle" (SRP) now attaches to
the emerging peptide chain and blocks further translation. When the
complex between SRP, nascent chain, and polysome makes contact
with a 72,000-dalton "docking protein" on the ER membrane, the
SRP is displaced and translation resumes. Since the process was
demonstrable in a cell-free reticulocyte system but not in a nuclease-
treated wheat germ system, an assay of reconstitution became available. An acidic ribonucleoprotein complex (with a molecular weight
of about 250,000) was isolated from reticulocyte lysate that blocks
translation of many nascent secretory proteins. Using a combination
of washing with salts and mild proteolysis, a reconstitution assay
was developed for both docking protein and SRP.

Another most useful approach to reconstitution from without was
devised by Schramm (1979a). Reconstitution of a functional hor-
mone-sensitive adenylate cyclase was achieved by fusion of two
defective membranes, each of them lacking a different step in the

hormone-mediated reaction. This complementation assay provided a tool for measuring the individual components of this complex receptor of the plasma membrane.

Enzymes such as cytochrome oxidase incorporated into liposomes have been reconstituted from without into erythrocytes by fusion (Gad *et al.*, 1979).

Another fascinating example involves the reconstitution of the Golgi apparatus (Rothman, 1981; Balch *et al.*, 1984). A crude membrane fraction prepared from a mutant cell infected with vesicular stomatitis virus was mixed with a comparable fraction from uninfected normal cells. The mutant membranes lacked a key glycosyltransferase and as a result did not catalyze the terminal stages of oligosaccharide processing and transport. Addition of normal membranes or even of Golgi membranes from rat liver, an unrelated tissue, restored the entire process by providing functional acceptor membranes. This intermembrane reconstitution only worked if ATP and soluble protein factors were added. More of this later.

What can we learn from such experiments on resolution and reconstitutions from without? The first and most obvious information is about the role of the protein and its subunits in the performance of the physiological function. In the case of the mitochondrial coupling factor (F_1), the association of an ATPase activity with a pure protein that restored oxidative phosphorylation pointed to a role of F_1 as a component of an ATP-driven ion pump.

Two other important pieces of information emerged from experiments on reconstitution from without. The first one was that the membrane altered the catalytic and physical properties of F_1 (Racker, 1963, 1976). The soluble enzyme was resistant to oligomycin and was cold-labile. The membrane-bound enzyme was sensitive and cold-stable. We speak of allotopic properties of the enzyme, but point out that the properties of the membrane were altered as well. Exposure to trypsin, which was well tolerated by the F_1 membrane complex, became very detrimental to the F_1-depleted membrane. Trypsin resistance was restored when F_1 was rebound to the membrane.

A second phenomenon became apparent from hybrid reconstitution experiments with submitochondrial particles and F_1 from either yeast or bovine heart mitochondria (Schatz *et al.*, 1967). In such reconstituted hybrid particles, only antibodies against homologous F_1, but not against heterologous F_1, inhibited ATP formation. This indicated that F_1 fulfilled two functions: a catalytic and a structural. To fulfill the structural function, heterologous F_1 was satisfactory. However, only homologous F_1 fulfilled the catalytic function as an energy transformer. Only yeast F_1 generated ATP in yeast particles; only bovine F_1 in bovine particles. Antibody against bovine F_1 inhibited only bovine particles; antibody against yeast F_1 only yeast particles. Yet under the experimental conditions designed, both yeast and bovine F_1 were required for either particle. We shall return to this problem when we discuss the function of the individual components of the mitochondrial proton pump.

Questions that can be answered by performing experiments on "reconstitution from without"
   1. What is the role of a protein in the function of a transporter?
   2. How does the interaction between the added protein and the other membrane proteins change the properties of each?
   3. Does the reconstituted protein have a structural as well as a catalytic function?

E. Reconstitution Starting with Intact Organelles or Vesicles

Before turning to the problem of resolution with detergents, I want to point out a fact which seems to be little appreciated. It is possible to reconstitute transport systems into phospholipid bilayers starting with intact organelles or membrane vesicles without the use of detergent. We stumbled upon this many years ago without quite knowing what we were doing (Conover *et al.*, 1963). We sonicated mitochondria in the presence of sonicated suspensions of crude soybean phospholipids and separated "P-particles." These vesicles are probably the first example of reconstituted proteoliposomes. They required F_1 and several other "coupling factors" for oxidative phosphorylation. The nucleotide and the P_i transporters of submitochondrial particles can also be directly incorporated into liposomes by the sonication method described by Shertzer *et al.* (1977). Another example is the reconstitution of ion transport systems starting with native sarcoplasmic reticulum vesicles. These systems can be incorporated by freeze–thaw sonication into liposomes that catalyze Ca^{2+} and P_i transport with properties characteristic of reconstituted vesicles (Carley and Racker, 1982). Since these properties are clearly different from those of the original organelles and since the morphological and sedimentation properties are also different, there can be little doubt that reconstitution into liposomes has been achieved.

One of the pitfalls of this approach is that in some instances it is difficult to be certain that the biological activity observed is not due to residual activity present in the original membrane vesicles. Thus, it is often advantageous to disrupt the original vesicles and destroy residual activity before embarking on reconstitution. This was done in the case of the reconstitution of the tetrodotoxin-sensitive Na^+ channel from lobster nerves (Villegas *et al.*, 1977).

The usefulness of the approach of reconstitution without resolution became apparent in more recent experiments on the reconstitution of the acetylcholine receptor. Early attempts to reconstitute purified preparations of acetylcholine receptor had met with little success in numerous laboratories, including our own. We therefore started with membrane vesicles that had not been previously exposed to a detergent (Sobel *et al.*, 1977). Having successfully converted these membranes into crude proteoliposomes (Epstein and Racker, 1978), we were equipped with a biological assay that allowed us to purify the receptor with retention of reconstitutive capacity and to characterize its functional properties after isolation (Huganir and Racker, 1982). It thus became possible to elucidate its requirements for lipids using pure lipids and phospholipids (Kilian *et al.*, 1980).

F. Resolution and Reconstitution of Membrane Complexes with Detergents

1. General Comments

Our work on resolution and reconstitution with detergents has been criticized. Mark Twain may have responded by saying that half of the lies that people spread about our reconstitution work are actually true.

a. The first criticism is that we are using a huge excess of phospholipids and are, therefore, dealing with artifacts. This is correct and we want it that way.

b. The second criticism is that we are using very crude preparations for reconstitutions. This is correct and we want it that way.

Let me explain. (a) Obviously, any removal of a protein from its natural membrane and separation from its natural neighbors is an artifact. We want it that way, because it allows us what no other method is capable of doing, namely, to resolve the system into individual components. Just like soluble enzymes have been isolated and analyzed in test tubes, membrane proteins have to be isolated and tested in liposomes. As Peter Hinkle has phrased it: "Liposomes are the test tubes for membrane proteins."

Now to the second criticism. (b) Why do we use crude membrane preparations for reconstitution? For the same reason that we are using crude homogenates when we embark on the purification of an enzyme. For the past 30 years the key research strategy of our laboratory has been to rely heavily on biological assays starting with crude preparations. Not just an assay based, for example, on oxidation or reduction catalyzed by a respiratory enzyme, but an assay of its function as a member of a biological process within the membrane catalyzing, e.g., oxidative phosphorylation or ion transport.

For reasons stated earlier it is important to develop as soon as possible such a biological assay in addition to, e.g., an affinity label assay during purification. I still recommend not to waste clean thinking on dirty systems and to prepare pure proteins for study. But I say now that I would rather have a membrane protein dirty than impotent, incapable of interaction with the membrane. It is easier to remove dirt from a protein than to cure its impotence. A radioactive affinity label has to be very tightly bound to the protein so that it will not dissociate during fractionation. If it is an irreversible inhibitor (e.g., NEM), we are dealing with a protein corpse. Although we can learn a great deal of important anatomy from a corpse, we want a live protein when we study its function. On the other hand, there are cross-linkers, modified ligands, and antibodies that can be reversibly dissociated. Moreover, affinity labels, e.g., radioactive α-bungarotoxin, can be invaluable for rapid assays on multiple samples during purification.

2. Choice of a Detergent

How do we solubilize membrane components? We can use both ionic or nonionic detergents. Cholate and deoxycholate have been the most useful ionic detergents; Triton X-100 and octylglucoside the most useful nonionic detergents. Ionic strength and pH may be critical. With cholate we usually use 1–2% in the presence of 0.4 M ammonium sulfate. When dealing with detergents, we must remember that we should establish optimal concentrations as well as an optimal detergent-to-protein ratio. I shall return to this problem when we discuss hydrophobic inhibitors of membrane-associated processes. Because of their high critical micellar concentration, octylglucoside and cholate are easily eliminated by dialysis or passage through Sephadex. Triton X-100 and Nonidet P-40 are more difficult to remove, but methods of separation are also available. These two nonionic detergents, particularly after sitting on the shelf for a few years, may contain peroxides that damage some proteins. Different commercial preparations vary enormously and should be analyzed for peroxides (e.g., by testing for oxidation of an SH compound). SDS at low concentration is sometimes useful, particularly in combination with a nonionic detergent. There are no general rules; in each case we have to determine which detergent is tolerated best by the particular protein.

New detergents are being introduced for the resolution and reconstitution of membrane proteins. "Chaps" has been successfully used for the solubilization of opiate receptors (Simonds *et al.*, 1980), "Zwittergent" for the proton pump of yeast plasma membranes (Malpartida and Serrano, 1981a,b), digitonin for the β-adrenergic receptor (Stiles *et al.*, 1983), and octylglucoside for many membrane proteins (Racker *et al.*, 1979), particularly bacterial proteins such as bacteriorhodopsin. It was successfully employed to solubilize the recalcitrant lactose carrier (Newman and Wilson, 1980) and other sugar transport systems in bacteria (Tsuchiya *et al.*, 1982). Other detergents with a high CMC, such as octylglucoside and octyl POE, have been used for the solubilization of membrane proteins and have been spectacularly successful for crystallization of hydrophobic proteins (Garavito and Rosenbusch, 1980; Michel, 1983).

The most direct approach to the problem of choosing a detergent for solubilization is to explore first which detergent is best tolerated by the membrane. More specifically, which detergent is best tolerated by the specific membrane function under investigation as monitored by the reconstitution assay. This is done by exposing the membrane to a variety of detergents both without and with added phospholipids, followed by the removal of the detergent by dialysis, a Sephadex column, or Bio-Beads. The preparation is then tested for residual activity (without added lipids) and for activity following reconstitution. When we used this procedure for the analysis of the carnitine transporter of submitochondrial particles (Schulz and

Racker, 1979), we found that the only detergent that did not seriously damage the transporter for subsequent reconstitution was octylglucoside. In the case of the P_i transporter of sarcoplasmic reticulum, the best detergent was cholate (Carley and Racker, 1982). For the acetylcholine receptor, octylglucoside was found to be very toxic, while cholate and Triton X-100 were well tolerated (Huganir and Racker, 1982).

Lesson 6: Choice of detergent

Don't *choose* a detergent. Try them all. Remember that nothing in reconstitutions is predictable.

Octylglucoside is the best for solubilizing lactose carrier; it is the best for bacteriorhodopsin; it damages the acetylcholine receptor.

Cholate is the best for the acetylcholine receptor, for cytochrome oxidase, and for the mitochondrial proton pump. It is very poor (if used alone) for the Ca^{2+}-ATPase and Na^+,K^+-ATPase.

Deoxycholate is the best for the Ca^{2+}-ATPase (together with cholate). It is bad for cytochrome oxidase.

Triton X-100 is good for the glucose transporter, for acetylcholine receptor; it is bad for the Na^+/Ca^{2+} antiporter.

Digitonin is the best for the adrenergic β-receptor.

Joy (DDAO) is good for bacteriorhodopsin, bad for the mitochondrial proton pump.

*Ivory Snow®** is best for diapers.

---

* A detergent, not cocaine.

Lecture 2

Methods of Resolution and Reconstitution

> Every part always has a tendency
> to reunite with its whole
> in order to escape its imperfection.
>
> **Leonardo da Vinci**

In the first lecture, I have given a broad overview of the approach of resolution and reconstitution. In this one I shall be dealing more with technical problems and specific examples.

I. Solubilization and Purification of Membrane Proteins

A. Gentle Procedures

Proteins can be removed from membranes with various degrees of brutality. Peripheral proteins may be removed by very gentle procedures such as exposure to hypotonic solutions, sometimes aided by low concentrations of a chelating agent (e.g., 0.5 mM EDTA). This method effectively removes chloroplast ATPase (CF_1) from chloroplasts (McCarty and Racker, 1966) or ATPase from *Streptococcus faecalis* membranes (Abrams, 1965) but is ineffective in the case of mitochondrial F_1. To remove F_1 from submitochondrial particles, shaking in the presence of glass beads (Pullman *et al.*, 1960) or vigorous sonication (Horstman and Racker, 1970) is required. Gentle hypotonic conditions were used many years ago to disrupt closed membranes and thus allow for the removal of cytochrome *c* from

18

mitochondria by extraction with 0.15 M KCl (Jacobs and Sanadi, 1960).

Exposure of erythrocyte ghosts to low ionic strength at an alkaline pH in the presence of a chelator resulted in the release of spectrin (Marchesi and Steers, 1968). Depending on the temperature, as much as 25% of the total ghost protein could be solubilized (Fairbanks *et al.*, 1971). Whereas spectrin was released in a few minutes at 37°C, several hours were required for its release at 0°C. Organic cations (e.g., tetramethylammonium bromide) greatly facilitated the release of proteins from human erythrocyte membranes. In fact, on prolonged incubation most of the membrane proteins were "solubilized" under these conditions (Reynolds and Trayer, 1971).

This raises the semantic question of what is a "peripheral protein." While there is an unambiguous answer in the case of F_1 or CF_1, which can be identified as peripheral proteins by electron microscopy, there is no clear answer in many other instances. For example, low concentrations of silicotungstate removed several coupling factors as well as succinate dehydrogenase from submitochondrial particles (Racker *et al.*, 1969). At least one of them (F_6) could not be released by conventional methods for removing peripheral proteins and required strong chaotropic agents for resolution (Knowles *et al.*, 1971).

Raising the ionic strength sometimes releases a peripheral protein. Acetylcholine esterase was released at high ionic strength from erythrocytes (Mitchell and Hanahan, 1966). However, high concentrations of salt sometimes release membrane proteins in an inactive form, an observation which has been successfully used to deplete membranes of a peripheral protein such as chloroplast CF_1 by treatment with high concentrations of sodium bromide (Kamienietzky and Nelson, 1975). Low concentrations of urea have been used to inactivate and remove a peripheral protein such as F_1 (Racker and Horstman, 1967).

Resolution of a membranous protein at an alkaline pH was successfully used for succinate dehydrogenase (King, 1963), mitochondrial ATPase inhibitor (Horstman and Racker, 1970), and other mitochondrial proteins, sometimes in combination with mild sonication (Racker, 1970). A particularly successful example of the use of a very alkaline pH to remove a membranous component is the case of the 43,000-dalton protein present in enriched membranes containing acetylcholine receptors. This procedure removed not only a protein that was not essential but also eliminated a controversy with respect to the possible function of this protein (Neubig *et al.*, 1979).

An unusual example for the "gentle" resolution of an active peripheral protein is the procedure developed for the isolation of F_1 (Beechey *et al.*, 1975). In this method, which was successfully adopted for chloroplasts (Younis *et al.*, 1977), the organelles were shaken briefly in the presence of chloroform resulting in the release of F_1 or CF_1.

Controlled exposure to proteases has been successfully used for the release of surface antigens (cf. Winzler, 1969). Hydrolases have been eluted after mild protease digestion of membranes (Schmidt and Thannhauser, 1943; Forstner, 1971) or by butanol extraction (Morton, 1955). Butanol alone may be ineffective, but effective if aided by the presence of urea (Kundig and Roseman, 1971). Solubilization by proteolysis or solvent extraction can introduce subtle changes into the released protein that may alter its biological properties. If, for example, proteolysis releases a protein because of induced changes in the attachment site on the membrane without altering the protein itself, the procedure is safe. If, however, the protein is released because it is severed from a covalently bonded hydrophobic segment, then the procedure yields an altered protein. Most revealing information can be obtained from such studies as in the case of cytochrome b_5 (Spatz and Strittmatter, 1971). Another example is the release of the docking protein from ER by mild proteolysis in the presence of salt and its reassociation after salt removal.

The great advantages of a gentle procedure of resolution is not only that the released protein is more likely to be in its native form but it also offers an assay system of reconstitution into the residual membrane fragments. That proved of crucial importance in the case of succinate dehydrogenase released at an alkaline pH, a procedure that did not damage the membrane. The enzyme was shown to be reconstitutively active after purification but only when isolated in the presence of succinate. When isolated in its absence it was fully active as a dehydrogenase but not capable of reconstitution with the membrane (King, 1963). Unfortunately other "gentle" procedures, such as shaking with chloroform or treatment with proteases, often (though not always) inactivate or severely damage the functional properties of the residual membrane.

B. Solubilization with Detergents, Solvents, or Chaotropic Agents

Procedures that solubilize an appreciable fraction of membrane proteins often destroy the biological function of the membrane and render the residual particles unsuitable for assay purposes (see below). Since most integral membrane proteins require relatively drastic methods for solubilization, there is really no choice. There are only a few rules regarding suitable methods for the solubilization of membrane proteins. Excellent reviews on the properties of detergents, their critical micellar concentration, and their use in membrane solubilization are available (Penefsky and Tzagoloff, 1971; Tzagoloff and Penefsky, 1971; Kagawa, 1972; Helenius and Simons, 1975; Tanford and Reynolds, 1976; Helenius et al., 1979; Grabo, 1982; Lichtenberg et al., 1983). I shall not discuss the properties of the various detergents that are described in these reviews. Instead, I shall attempt to extract from these reviews data and observations that may help us in a more rational approach to the resolution and reconstitution of membrane proteins. The critical micellar concentration (CMC) and the hydrophil–lipophil balance (HLB), the binding properties to various proteins, the critical temperature of crystallization and of demixing, and the micellar size all affect the solubilization properties of detergents (cf. Grabo, 1982). Nonionic detergents with an HLB in the range of 12.5–14.5 that are among the most effective include Triton X-100, Nonidet P-40, and Brij 56. Detergents with higher HLB. e.g., the Tweens or Brij 58, release peripheral proteins. To extract integral protein or lipids, a high pH or repeated extraction is required with these detergents. Of particular importance is the CMC of detergents of which a few are listed in Table 2-1. Both temperature and ionic strength and presence of impurities greatly influence the CMC and the solubilization of hydrophobic proteins. It is therefore not surprising that procedures of purification are often difficult to reproduce in other laboratories (Lesson 7). Ionic detergents, such as the bile salts, which contain carboxyl groups are very dependent on pH and should not be used below 7.0. Conjugated bile salts and zwitterionic detergents do not have this disadvantage. Sodium dodecylsulfate is an excellent solubilizer of membrane proteins but is usually very damaging. Nevertheless, sodium dodecylsulfate is being used with remarkable suc-

TABLE 2-1 Critical Micellar Concentration of Some Detergents Commonly Used for Resolution-Reconstitution Experiments

Detergent	CMC (mM)
Triton X-100	0.24
Sodium cholate	13–15
Sodium deoxycholate	4–6
Lysolecithin mixture	<0.10
Octylglucoside	25
Octyl POE	6.6
$C_{12}E_8$	0.087

cess for the purification of the Na^+,K^+-ATPase from plasma membranes (Jørgensen, 1975). In this case low concentrations of detergent were used to extract "impurities" leaving the active Na^+,K^+-ATPase behind. A mixture of ionic (cholate) and nonionic (octylglucoside) detergent was found to be particularly suitable for the extraction of the chloroplast proton pump (Pick and Racker, 1979a). In contrast to Triton X-100 and NP-40, octylglucoside and octyl POE, which have a high CMC, are more readily removed by dialysis. As mentioned earlier, Triton X-100 accumulates peroxides on aging and becomes increasingly toxic. Methods for removal of peroxides have been described (Ashani and Catravas, 1980), but it is simpler and cheaper to buy a new bottle and analyze it for peroxides (e.g., by measuring the oxidation of DTT). Nonionic detergents can be removed with Bio-Beads or Extracti-Gel D, but proteins are sometimes removed as well. An alternative procedure to remove Triton X-100 or NP-40 is to exchange it against ionic detergents followed by precipitation of the protein (Hoffman, 1979; Huganir and Racker, 1982).

Lesson 7: On the reproducibility of purification procedures

If you cannot repeat a procedure of purification described by someone else, look for other reasons before you denounce that investigator as incompetent. Try once to repeat one of your own procedures, e.g., for the purification of a yeast enzyme, 5 years after you published it. You will be surprised how a different batch of cholate (which you didn't recrystallize) can change the results. What about the yeast you grow or buy? A well-known supplier of crystalline enzymes had to stop delivery of a most useful enzyme because the import of Castro's sugar had ceased. Another source of sucrose gave a completely altered enzyme fractionation pattern. Do you use Tris or cholate or deoxycholate? Did you notice how some batches are yellow in concentrated solution? We buy large batches of cholate and deoxycholate and crystallize them. By the time we have used them up we have published the paper. Do you check how much acid there is in your glycerol? We examined 8 bottles of Triton X-100 and NP-40 that had been stored in various laboratores in our department and found staggering differences in contaminations with peroxides. We buy small bottles of Triton X-100 and NP-40 and keep the solutions under nitrogen to prevent peroxide formation. I think we should describe purification procedures as follows: Use cholate to extract the enzyme. Try 0.2–0.4 mg of cholate per mg of protein in the presence of 0.5–1.0 M KCl. find out what works. Unfortunately, unreasonable editors (including myself) insist on an accuracy that has little meaning.

If a membranous protein is found to be unstable in the presence of detergents, many common stabilizers should be tried: cofactors and substrates (or reversible inhibitors), reducing agents (dithiothreitol, monothioglycerol, or mercaptoethanol), chelating agents (EDTA), protease inhibitors, antifreeze agents (glycerol, sucrose, or ethylene glycol), and above all, phospholipids.

The sequence of a recommended approach to reconstitution is as follows: (a) development of a functional assay, starting with membrane fragments, (b) exploration of single and multiple detergents for stability of function followed, if necessary, by testing of stability in the absence and presence of stabilizers, and (c) solubilization with single or multiple detergents that are best tolerated by the protein.

Lesson 8: The protein

Some membrane proteins are tough. If you happen to drop a bottle of bacteriorhodopsin on the floor, don't despair. I have seen a student using distilled water to collect it from the floor, which is unnecessary sophistication. In any case you will find the protein is still reconstitutively active. It may, of course, have gained radioactivity.

Some membrane proteins are much more delicate. Wrap them up with phospholipids, coat them with glycerol, shield them with dithiothreitol of (if your budget is low) with monothioglycerol. Remember to check a fragile protein for its pH preference. A few like it acid, some like it basic, though most are partial to neutrality. Few, but very few, like it better hot (20°C) than cold (0°C).

II. Purification of Membrane Proteins

Once the protein is solubilized, it may or may not depend on the presence of a detergent to remain in solution. How can a hydrophobic protein be soluble in an aqueous medium without the aid of a detergent? There are two explanations for this well-established fact. The first one is that even after removal there is sufficient residual detergent bound to the protein to keep it in solution. This appears to be the case with cytochrome oxidase, which does not precipitate out until the cholate/protein ratio drops below 100 μg/mg (Eytan *et al.*, 1976). A second and more intriguing explanation has emerged from studies of proteolipids (Folch-Pi and Stoffyn, 1972). These remarkable proteins are soluble in chloroform–methanol (2:1) but are readily converted into water-soluble proteins by the slow exchange of the organic solvent with water. These opportunistic chameleonlike proteins can expose either their hydrophobic or hydrophilic amino acid residues depending on the environment. They represent a good illustration for the semantic difficulties in defining a hydrophobic protein by its ratio of hydrophobic and hydrophilic amino acid residues.

Lesson 9: What is a hydrophobic protein? Can we purify it by conventional methods?

Is a proteolipid a hydrophobic or a hydrophilic protein? This is as difficult to define as the position of some politicians. When they face Arabs they are Arabs, when they face Jews they are pro-Israel. A proteolipid facing water is exposing its hydrophilic residues while hiding its hydrophobic residues. When facing the phospholipid bilayer after incorporation into a membrane (e.g., of mitochondria), it is hydrophobic, exposing a large amount of hydrophobic amino acid residues, but can form a channel which is hydrophilic on the inside, thereby fulfilling its physiological function. In any case, insolubility in water is a poor criterion for definition or we would have to call denatured serum albumin a "hydrophobic" protein. It is important to realize that a hydrophobic protein may be soluble in aqueous solutions without or with a minimal amount of detergent and thus be amenable to purification by conventional methods designed for hydrophilic proteins.

Thus, the purification of membrane proteins can often be approached by conventional methods, such as salting out, ion exchange, exclusion, and affinity chromatography. However, if the protein shows a tendency to aggregate, either detergents or phospholipids must be included in each step of purification. Fractionation with ammonium sulfate in the presence of a detergent such as cholate has been particularly useful with membrane proteins from mitochondria and chloroplasts as well as with membrane proteins extracted from plasma membranes with neutral detergents such as NP-40. If a fraction prefers to float rather than sediment, we proceed with its separation by removal of the subnatant. Sometimes one can remove the salt by dialysis against sucrose or glycerol without precipitation of the protein. If a detergent is required to prevent aggregation, usually a much lower concentration than needed for extraction may suffice to keep the protein in solution. Hydrophilic columns such as hydroxylapatite or DEAE-cellulose can then be used in the traditional manner to adsorb the proteins and to elute them at increasing salt concentrations in the presence of detergent. Hydrophobic columns such as phenyl-Sepharose can be used to adsorb membranous proteins at high salt and to elute them with a decreasing salt gradient. In either case the presence of a detergent may be required. Investment in the preparation of an affinity column is very much worthwhile. It is a marvelous tool for the simultaneous purification and concentration of membrane proteins. Other concentration methods include ion exchange columns, dialysis against solid sucrose or 30% polyethylene glycol, or filtration through membranes with appropriate cut-off sizes. Lyophilization is rarely useful with membrane proteins unless sucrose or glycerol is included during the process to avoid denaturation.

Particularly valuable are the recently developed high-performance exclusion, ion exchange, and hydrophobic columns that allow rapid separation of proteins. Detergents or phospholipids may again be needed to recover the protein in acceptable yields. Sucrose gradient fractionations, isoelectric focusing, chromatofocusing, and dye columns with selective absorption capacity are often effective for purification of hydrophobic proteins and can be used in the presence of low concentrations of detergents. The instruction booklets issued by the manufacturers are supplied without charge together with the overpriced column materials.

Lesson 10: Evolution of lessons for purification

1963 Don't think, purify first.
Don't waste clean thinking on dirty enzymes.

1973 Don't purify more than necessary. A dirty protein is better than an impotent one. Tomorrow, the dirty protein may be clean, but the dead enzyme will still be dead (unless resurrected by Khorana*).

1980 If the purified, reconstitutively active receptor shows a single band in SDS–PAGE, it means either (a) you got it, or (b) a proteinase has nicked it. Acetylcholine receptor nicked by pronase shows two bands on SDS–PAGE (23K and 10K), yet has a molecular weight of 250,000 in a sucrose gradient and is reconstitutively active. CF_1 with 2 subunits is much cleaner and more active as ATPase than CF_1 with 5 subunits, but is much deader in reconstitution.

1984 Don't think, don't purify: clone.

III. Methods of Reconstitution

There are basically three different methods of reconstitution of membrane proteins.

1. *Reconstitution from without.* In this method, mentioned in the first lecture, a purified membrane protein, which may be peripheral or integral, is restored to a membrane. The membrane may be the outer or inner face of an organelle, it may be the plasma membrane

* See Lecture 8.

of an intact cell, or it may be a phospholipid bilayer. The latter may be a liposome, a proteoliposome, or a flat membrane threatened on each side by electrodes of an electrophysiological biochemist. The recombination process may be spontaneous or aided by mild detergents or fusogens. For incorporation into planar membranes, the protein may be naked or first reconstituted into liposomes. Excellent reviews on some of these procedures have been published (Mueller, 1975; Eytan, 1982; Montal *et al.*, 1981; Miller, 1983; Klausner *et al.*, 1984).

2. *Reconstitution by sonication*. In this method the membrane structure is lost and regained. No detergents are used, and the most commonly used method of disruption is oscillation in a bath-type sonicator. The reconstitution partners are mixed and either sonicated directly (Racker, 1973) or exposed prior to sonication to freeze–thawing (Kasahara and Hinkle, 1976) or to a chemical such as bromohexane (Racker *et al.*, 1979).

3. *Reconstitution of detergent-solubilized proteins*. The third method depends on the presence of a detergent at a concentration which disrupts the membrane structure of the phospholipid bilayer. The detergent is subsequently removed by dialysis, dilution, exclusion chromatography, Bio-Beads, or Extracti-Gel D.

A. Reconstitution from Without

1. Spontaneous Reconstitution

a. *Peripheral Proteins*. In this method the active proteins are simply added to the depleted membrane. Time, temperature, and concentrations are important factors, as in all other processes of interaction. In the case of the inner mitochondrial membrane, the more depleted the membrane of coupling factors, the longer it takes to reconstitute the system. Particles only partially depleted of F_1 required 2–3 min for reconstitution with F_1 (provided a suitable concentration of F_1 was present). Particles more extensively depleted of F_1 as well as of other coupling factors by exposure, e.g., to silicotungstate, required a 30 min incubation with the coupling factors (Racker *et al.*, 1969). The order of the addition did not seem critical in this particular case but might be important in other instances. Cytochrome b_5 was shown to enter spontaneously into liposomes (Enoch *et al.*, 1977). A cation such as Mg^{2+} may be required, as in the case of F_1 (Pullman *et al.*, 1960), or of bacterial porin (Nakae, 1976). Sometimes association of peripheral proteins with the membrane is not specific and may be aided by salts. In this case a simple binding assay is misleading. For example, in the case of mitochondrial F_1, the binding to the particles in the presence of salt yielded a rutamycin-insensitive complex. It was only when a second coupling factor was present that a rutamycin-sensitive complex was formed (Bulos and Racker, 1968). Some recent studies on the binding of ATPase to depleted or mutant *Escherichia coli* membranes revealed a similar picture (Perlin *et al.*, 1983) and emphasize the pitfalls when reliance is placed on binding assays only. There are many examples for the requirement of Ca^{2+} for binding, sometimes aided by calmodulin (Cheung, 1980) or synexin (Creutz *et al.*, 1978; Hong *et al.*, 1981). Binding of thyrotropin was shown to require the presence of brain gangliosides (Aloj *et al.*, 1977); so does tetanus toxin (Borochov-Neori *et al.*, 1984) and Sendai virus (Haywood, 1983). In the case of the binding and transport of the riboflavin-binding protein, the state of phosphorylation of the protein was shown to be critical. The dephosphorylated protein was inactive when tested for transport across the membrane (Miller *et al.*, 1982).

b. *Integral Proteins.* The process of the spontaneous incorporation of integral proteins into preformed proteoliposomes (Eytan *et al.*, 1976; Eytan and Racker, 1977) is still mysterious. Why does cytochrome oxidase insert itself into liposomes and why does it do it unidirectionally? Why does the presence of one protein in the liposomes influence the incorporation of a second one? Why does the resident protein sometimes aid, sometimes hinder, the second invader? Do we deal here with a primitive case of molecular psychology? We shall return to the problem of unidirectional orientation after the discussion of other reconstitution procedures. The key feature for the successful spontaneous incorporation of proteins into liposomes is the presence of acidic phospholipids in rather high proportions (10–30%). The process is rapid, requiring only 2–3 min at room temperature in the presence of Mg^{2+} and takes about 5–8 min in the absence of Mg^{2+} and in the presence of EDTA. It is considerably slower at lower temperatures. Low concentrations of detergents, such as cholate or lysolecithin, inhibit the incorporation. There are limited comparative data for the incorporation procedure.

TABLE 2-2 **Activities of Systems Reconstituted by the Spontaneous Incorporation Procedure**

Reconstituted system	Activity[a]
$^{32}P_i$–ATP exchange of F_1F_0	60 nmol $\times$ min^{-1} $\times$ mg protein^{-1}
Cytochrome oxidase	Respiratory control ratio of 6–10
QH$_2$–cytochrome c reductase	Respiratory control ratio of 5–6
Light-driven ATP generation with bacteriorhodopsin and ATPase	30 nmol $\times$ min^{-1} $\times$ mg protein^{-1}

[a] All values are for activities at 25°C (or room temperature).

It should be emphasized that the method as we used it was not successful with some proteins, such as bacteriorhodopsin and Na$^+$,K$^+$-ATPase. Some successful examples are listed in Table 2-2.

2. Reconstitution by Facilitated Incorporation

a. *By Detergents.* This procedure was described by Racker (1972a) and Eytan *et al.* (1975). Preformed liposomes are exposed to dilute solutions of membrane proteins at 0°C for several hours in the presence of very small amounts of detergent. This method was originally devised in an attempt to obtain insight into the process of membrane biogenesis. Lysolecithin was therefore used as a preferential detergent. However, cholate at low concentration (0.1%) was equally effective in some systems. Liposomes made with soybean phospholipids and 10% lysolecithin were used in the experiments recorded in Table 2-3.

TABLE 2-3 **Activities of Systems Reconstituted by Facilitated Incorporation Procedure**

Reconstituted system	Activity[a]
$^{32}P_i$–ATP exchange of F_1F_0	50–70 nmol $\times$ min^{-1} $\times$ mg protein^{-1}
Cytochrome oxidase vesicles	Respiratory control ratio of 4–10
Coenzyme Q–cytochrome c reductase	Respiratory control ratio of 3–5
Ca^{2+} pump of sarcoplasmic reticulum	0.1–0.2 μmol $\times$ min^{-1} $\times$ mg protein^{-1}

[a] All values are for activities at 25°C (or room temperature).

b. *By Fusion*. Liposomes that contain phosphatidylethanolamine and about 30% of either phosphatidylserine or cardiolipin (but not phosphatidylinositol) fuse rapidly on addition of Ca^{2+}. The fusion procedure has particular value for the reconstitution of mixtures of proteins that are optimally reconstituted by different procedures. For example, cytochrome oxidase reconstituted into liposomes by the cholate-dialysis or incorporation procedure, exhibiting oxidase activity with a high respiratory control, can be fused with vesicles that contain the hydrophobic proteolipid of mitochondria reconstituted by the sonication procedure (Miller and Racker, 1976a). Fusion was measured kinetically by the increase in respiration (loss of respiratory control) when the proton channel of the oligomycin-sensitive ATPase was incorporated by fusion into the same vesicle that contained cytochrome oxidase.

Another advantage of the fusion procedure is that it yields larger liposomes than the other reconstitution procedures. By the use of osmotic gradients, fusion between large vesicles and even with planar phospholipid bilayers has been achieved (Miller *et al.*, 1976; Miller and Racker, 1976b). Preparation of giant liposomes by fusion with Sendai virus or fatty acids will be discussed in Lecture 12.

An important, more recent development has been the use of fusion of liposomes with cell membranes, intact cells, or organelles. For example, highly purified integral proteins, such as cytochrome oxidase, ATPase of bovine heart mitochondria, or Ca^{2+}-ATPase of sarcoplasmic reticulum have been reconstituted into liposomes and then incorporated into erythrocytes by fusion in the presence of Ca^{2+} (Eytan and Eytan, 1980; Gad *et al.*, 1979). Ingenious new fusion methods have been developed to analyze complex receptor systems. For example, the β-adrenergic receptor of turkey erythrocytes was incorporated by fusion, with the aid of Sendai virus, and coupled to the adenylate cyclase of adrenal tumor cells (Schramm, 1979a). A more efficient and more generally applicable method of fusion with polyethylene glycol, phospholipids, and ATP was used to incorporate the glucagon receptor from liver membranes into the plasma membrane of Friend erythroleukemia cells (Schramm, 1979b). These methods were then extended to allow for the quantitative analysis of a functional hormone receptor, guanyl nucleotide binding protein, and the catalytic subunit of the adenylate cyclase (Neufeld *et al.*, 1980). Using a similar procedure, the reconstituted asialoglycoprotein from mouse hepatocytes was fused in the presence of polyethylene glycol with mouse fibroblasts, which acquired the ability to degrade the glycoprotein (Bauman *et al.*, 1980). The erythrocyte anion exchange protein was reconstituted with the envelope proteins of Sendai virus, allowing the fusion into cells that were deficient in anion exchange, and thus became competent (Volsky *et al.*, 1980). Fusion of liposomes with mitochondria has been described (Schneider *et al.*, 1980). The method should be useful for reconstitutive complementation assays with defective mutant mitochondria.

The use of fusion for the incorporation of proteoliposomes into planar bilayer membranes has been extensively reviewed (Miller and Racker, 1979; Montal *et al.*, 1981; Miller, 1983).

c. *By Freeze–Thawing.* This procedure was first used for the reconstitution of the chloroplast proton pump ATPase (Pick and Racker, 1979a). It is very suitable for the formation of giant proteoliposomes and was used successfully for the reconstitution of the amino acid transporter from Ehrlich ascites tumor cells (McCormick *et al.*, 1984; M. Schenerman and M. Newman, unpublished experiments, 1984) and lactate transporter (M. Newman, S. Nisco, and E. Racker, unpublished experiments, 1984) from rabbit erythrocytes. These vesicles have a large trapping volume and are, therefore, particularly suitable for the study of transport of solutes such as lactate that give rise to relatively high leakage rates in unilamellar liposomes in the absence of protein. Even short exposure to sonication greatly reduces their trapping volume and transport activity. They must be formed in the absence of sucrose or glycerol.

B. Reconstitution by Sonication

Technical details regarding the type of sonicator, time, and temperature of exposure have been previously described (Racker, 1979). Probably the most important aspect of this procedure is that no detergents are used. As mentioned earlier, organelles (e.g., sarcoplasmic reticulum) and even intact cells (e.g., *Halobacterium halobium*) can be directly incorporated into liposomes. Reconstitution was readily established in the case of *H. halobium* because the direction of proton pumping reversed on reconstitution (E. Racker, unpublished observation). Reconstitution with crude membranes is of particular value for establishing an assay method before a choice of detergent for solubilization is made.

1. Direct Sonication

This procedure was first developed for the reconstitution of bacteriorhodopsin (Racker, 1973). It yielded proteoliposomes with a proton pumping activity much higher than that obtained by cholate dialysis, but was not as suitable for coreconstitution with the mitochondrial ATPase (Racker and Stoeckenius, 1974).

As can be seen from Table 2-4, the rates of Ca^{2+}-translocation by proteoliposomes obtained by sonication were about one half of those observed with vesicles obtained by cholate–deoxycholate dialysis, but manipulation of the phospholipid composition (Racker and Eytan, 1973) yielded vesicles with Ca^{2+}-translocation rates approaching those obtained by the dialysis procedure.

2. Freeze–Thaw Sonication

This procedure was first used for the reconstitution of the glucose transporter of human erythrocytes (Kasahara and Hinkle, 1976). Liposomes were prepared by sonication and were added to the protein. The mixture was then quickly frozen by immersing the test tube into a dry ice–acetone bath or into liquid nitrogen. After thaw-

TABLE 2-4 **Activities of Systems Reconstituted by the Direct Sonication Procedure**

Reconstituted system	Activity[a]
$^{32}P_i$–ATP exchange of F_1F_0	50–80 nmol $\times$ min^{-1} $\times$ mg protein^{-1}
Oxidative phosphorylation site II plus III	P/O ratio of 0.2–0.4
Oxidative phosphorylation site III	P/O ratio of 0.2–0.4
Ca^{2+} pump of sarcoplasmic reticulum	0.3–0.4 μmol $\times$ min^{-1} $\times$ mg protein^{-1}
Cytochrome oxidase vesicles	Respiratory control ratio of 3–5
Proton pump of *Halobacterium halobium*	2–6 μmol $\times$ min^{-1} $\times$ mg protein^{-1}
Na^+ pump of electric eel ATPase	0.2–0.3 μmol $\times$ min^{-1} $\times$ mg protein^{-1} (at 30°C)
Adenine nucleotide transporter	0.6–0.8 μmol $\times$ min^{-1} $\times$ mg protein^{-1}
P_i transporter of mitochondria	0.6–0.8 μmol $\times$ min^{-1} $\times$ mg protein^{-1}

[a] All values are for activities at 25°C (or room temperature).

ing at room temperature, the mixture was exposed to brief periods of sonication, usually for less than 2 min. In the case of the glucose transporter, optimal sonication time was 20 sec; for the Na^+, K^+-ATPase of electric eel we observed it to be 1 min; and for bacteriorhodopsin about 2 min. The exact time for optimal results varies with the volume, geometry of the test tube, power output, etc. The greatest advantage is that the method can be used for proteins that are sensitive to prolonged sonication, such as cytochrome oxidase, without much loss of enzymatic activity. In the case of the Na^+, K^+-ATPase of the electric eel, the optimal time by direct soniciation was 30 min as compared to 1 min by freeze–thaw sonication, and the rates of Na^+ transport by the latter procedure were significantly higher (D. Cohn and E. Racker, unpublished observations, 1977). The highest rates obtained by direct sonication were about 300 nmol $\times$ min^{-1} $\times$ mg protein^{-1} (at 30°C), whereas rates of about 1000 nmol (at 37°C) were obtained by the freeze–thaw procedure. In the case of reconstitution of bacteriorhodopsin with the mitochondrial ATPase, the rates achieved by the two procedures were similar. Some of the rates obtained with proteoliposomes reconstitututed by freeze–thaw sonication are given in Table 2-5.

A useful variant of the freeze–thaw sonication procedure is the inclusion of detergents. In some systems the presence of cholate greatly facilitates the formation of active proteoliposomes. This

TABLE 2-5 **Activities of Systems Reconstituted by the Freeze–Thaw Sonication Procedure**

Reconstituted system	Activity[a]
$^{32}P_i$–ATP exchange of F_1F_0	70 nmol $\times$ min^{-1} $\times$ mg protein^{-1}
Cytochrome oxidase	Respiratory control ratio of 4
Ca^{2+} pump of sarcoplasmic reticulum	0.8 μmol $\times$ min^{-1} $\times$ mg protein^{-1}
Proton pump of *Halobacterium halobium*	4 μmol $\times$ min^{-1} $\times$ mg protein^{-1}
Photophosphorylation with *Halobacterium halobium* and mitochondrial ATPase	50 nmol $\times$ min^{-1} $\times$ mg protein^{-1}
Na^+ pump of electric eel ATPase	0.8–1.2 μmol $\times$ min^{-1} $\times$ mg protein^{-1} (at 37°C)

[a] Unless specified otherwise, all values are for activities at 25°C (or room temperature).

method was used for the incorporation of F_0 into bacteriorhodopsin vesicles, followed by reconstitution with coupling factors (Alfonso *et al.*, 1981), for the reconstitution of Na^+K^+ pumps (Karlish and Pick, 1981), and recently for the reconstitution of the anion transporter of erythrocytes (I. Ducis, A. Kandrach, and E. Racker, unpublished observations) as well as for the amino acid transporter (Schenerman *et al.*, 1985). The optimal cholate concentration, as well as cholate-to-protein ratio, must be established for each system.

One complication with the freeze–thaw sonication procedure is that it fails in the presence of sucrose, glycerol, or other antifreeze compounds. This requires removal of sucrose, which is widely used for the stabilization of membranes during freezing for storage. Low concentrations of sucrose during freeze–thaw may be tolerated, but usually require the presence of higher concentrations of salts. Both ionic strength and pH should be controlled for reproducibility. The effect of ionic strength, sucrose, and phospholipid concentration on the freeze–thaw sonication procedure was described in detail (Pick, 1981).

An alternative to the freeze–thaw sonication procedure is the inclusion of some fusogens (e.g., bromohexane) during sonication (Racker *et al.*, 1979). The procedure worked well with cytochrome oxidase, but was unsuitable for bacteriorhodopsin. Moreover, scrambling (right side out and inside out vesicles) was encountered in contrast to the cholate dialysis procedure, which yields at the appropriate protein/phospholipid ratio exclusively unidirectional (right side out) vesicles.

C. Reconstitution of Detergent-Solubilized Proteins

As mentioned in the first lecture, the choice of a suitable detergent for the solubilization of a membrane protein is empirical and unpredictable. So is the choice of the method for the removal of the detergent. What needs to be emphasized is that the optimal time of exposure to the detergent as well as the optimal time allowed for its removal varies greatly with different proteins.

1. Detergent Dialysis (or Removal) Procedure

Solubilization with cholate and removal by dialysis was the first procedure used for the reconstitution of the mitochondrial proton pump (Arion and Racker, 1970; Kagawa and Racker, 1971), cyto-chrome oxidase (Hinkle *et al.*, 1972), and oxidative phosphorylation (Racker and Kandrach, 1971, 1973). The basic procedure is to mix a suspension of sonicated phospholipids with the isolated protein complex in the presence of appropriate concentrations of cholate (1–3%), followed by removal of the detergent by dialysis for 20 hr or longer. Sometimes cholate alone is not effective and mixtures of detergents are required. In the case of the sarcoplasmic reticulum Ca^{2+} pump, a mixture of cholate and deoxycholate gives optimal reconstitution (Knowles and Racker, 1975a). Similar observations were made with the reconstitution of the DCCD-sensitive proton pump from thermophilic bacteria (Sone *et al.*, 1975) and with the reconstitution of CF_1F_0 (Schmidt and Gräber, 1985). The latter pro-cedure, which was reproduced in our laboratory, gives remarkably high values for ATP formation. Cholate dialysis has been used early for reconstitution of the Na^+,K^+-ATPase (Goldin and Tong, 1974; Hilden *et al.*, 1974), but in our hands proved to be inferior to sonica-tion (Racker and Fisher, 1975). In any case, better methods for the reconstitution of this pump are now available and will be described later.

The major disadvantage of the dialysis procedure is the prolonged exposure to detergents, which is not well tolerated by some mem-brane proteins. Sometimes addition of dithiothreitol or methanol to the dialysis fluid helped to stabilize the protein (Kagawa and Racker, 1971). At least part of the need for a large excess of phospholipids for reconstitution is protection against the detergent. In fact, the phospholipid/protein ratio could be reduced by a factor of 5 when the exposure to detergent was shortened (Carroll and Racker, 1977). More rapid methods for the dialysis of cholate have been described (Sweadner and Goldin, 1975).

Another disadvantage of the dialysis procedure is the time factor. During fractionation of membrane proteins a simple and rapid method of analysis is desirable, particularly if the protein is labile.

We have used a Sephadex column equilibrated with a low concen-tration of cholate for reconstitution of the mitochondrial proton pump (Kagawa *et al.*, 1973a). Too rapid removal through a deter-gent-free Sephadex column yielded an inactive preparation. When rapid removal is tolerated, Sephadex columns are very convenient. For assay of the proton pump, the detergent dilution procedure described in the next section is rapid and accurate.

TABLE 2-6 **Activities of Systems Reconstituted by the Detergent Dialysis Procedure**

Reconstituted system	Detergent	Activity[a]
$^{32}P_i$–ATP exchange of F_1F_0	Cholate	100–150 nmol $\times$ min^{-1} $\times$ mg protein^{-1}
Oxidative phosphorylation site I	Cholate	P/O ratio of 0.3–0.5
Oxidative phosphorylation site II plus III	Cholate	P/O ratio of 0.4–0.6
Oxidative phosphorylation site III	Cholate	P/O ratio of 0.3–0.5
Ca^{2+} pump of sarcoplasmic reticulum	Cholate + deoxycholate	0.6–0.8 μmol $\times$ min^{-1} $\times$ mg protein^{-1}
Cytochrome oxidase vesicles	Cholate	Respiratory control ratio of 8–12
Na$^+$ pump of ATPase from *Squalus acanthias*	Cholate	5–12 nmol $\times$ min^{-1} $\times$ mg protein^{-1}

[a] All values are for activities at 25°C (or room temperature).

Some selected data obtained by the detergent dialysis procedure are summarized in Table 2-6.

2. The Detergent Dilution Procedure

This is by far the simplest and most rapid procedure, particularly when a sensitive assay system is available. The detergent dilution method was first used with cholate for the reconstitution of cyto-chrome oxidase (Racker, 1972a) and extended to several other membrane proteins. It is particularly suitable for the reconstitution of ATP-driven proton pumps (Racker *et al.*, 1975; Winget *et al.*, 1977). It consists of exposing sonicated phospholipids and the membrane protein to, for example, cholate (0.5–2.0%) for various periods of time followed by dilution into the assay mixture. This is possible when a sensitive assay is available that allows at least a 25-fold dilution of the reconstituted vesicles, thereby reducing the detergent to concentrations that do not interfere with the assay. When the assay is affected, the diluted vesicles can be concentrated by centrif-ugation (Miyamoto and Racker, 1980) or by concentration on a Milli-pore filter, which can be used as a support during assay (Newman *et al.*, 1981). Optimal dilution varies considerably with different membrane systems. A rather sharp optimum at 1:6 dilution of cholate was observed in the case of the Na$^+$/Ca^{2+} antiporter (Miyamoto and Racker, 1980). Particularly successful reconstitutions have been achieved by the octylglucoside dilution procedure. For example, in the case of bacteriorhodopsin, it was far better than sonication procedures (Racker *et al.,* 1979), wherease cholate dilution was completely ineffective. In contrast, in the case of the acetylcholine receptor, octylglucoside dilution was ineffective, whereas cholate dilution was moderately effective. Octylglucoside dilution has been particularly useful as the method of choice in the reconstitution of the lactose transporter (Newman and Wilson, 1980).

The only serious disadvantage of the detergent dilution procedure is that sometimes it does not work at all or yields lower activities than the dialysis procedure (e.g., in the case of the reconstitution of the acetylcholine receptor).

Data obtained with some membrane proteins by the detergent dilution procedure are listed in Table 2-7.

A curious aspect of the detergent dilution method is the time of contact with the detergent required before dilution. In the case of bacteriorhodopsin, exposure to 1.25% octylglucoside for 3 min was sufficient to obtain maximal pump activity. In the case of cytochrome oxidase, Roger Carroll, a graduate student at Cornell Uni-

TABLE 2-7 **Activities of Systems Reconstituted by the Detergent Dilution Procedure**

Reconstituted system	Detergent	Activity[a]
$^{32}P_i$–ATP exchange of F_1F_0	Cholate	120–160 nmol $\times$ min^{-1} $\times$ mg protein^{-1}
$^{32}P_i$–ATP exchange of CF_1F_0	Cholate	200–400 nmol $\times$ min^{-1} $\times$ mg protein^{-1}
	Octylglucoside	160–180 nmol $\times$ min^{-1} $\times$ mg protein^{-1}
Oxidative phosphorylation site III	Cholate	P/O ratio of 0.2–0.4
Ca^{2+} pump of sarcoplasmic reticulum	Cholate	400–600 nmol $\times$ min^{-1} $\times$ mg protein^{-1}
Ca^{2+} pump of sarcoplasmic reticulum	Octylglucoside	200–250 nmol $\times$ min^{-1} $\times$ mg protein
Cytochrome oxidase	Cholate	RCR[b] of 5–6
Cytochrome oxidase	Octylglucoside	RCR of 5–6
Na^+K^+ pump of electric eel ATPase	Octylglucoside	800 nmol $\times$ min^{-1} $\times$ mg protein^{-1}
Ca^{2+}/Na^+ antiporter	Cholate	30–40 nmol $\times$ min^{-1} $\times$ mg protein

[a] All values for activities at room temperature.

[b] RCR = respiratory control ratio.

versity, observed respiratory control ratios of 4 or greater after exposing the enzyme for 10–16 min to 1% cholate in the presence of purified mitochondrial phospholipids. After we had published this procedure (Carroll and Racker, 1977), I confidently tried it one day for the reconstitution of cytochrome oxidase with crude soybean phospholipids and failed. By that time Roger had received his Ph.D. degree and was far away (in California). Fortunately I was able to repeat his experiments with purified mitochondrial phospholipids. Since I wanted to coreconstitute cytochrome oxidase with a second protein, which worked best with crude soybean lipids, I reexamined the system. It turned out that the cholate dilution procedure could be used with soybean phospholipids, provided the incubation in the presence of 1% cholate was extended to 8–12 hr. This was rather surprising. We had previously assumed that the reason for the long time of dialysis needed for reconstitution of the mitochondrial proton pump (Kagawa and Racker, 1971) was to achieve the effective removal of the detergent. It now appears that there may be a second and unexplained reason for the need for a prolonged contact between the detergent and its reconstitution victim.

Lesson 11: The methods of reconstitution

If you want to reconstitute a membrane function, try it first with *crude membrane* fragments using:
1. Cholate dialysis
2. Freeze–thaw sonciation
3. Octylglucoside dilution

If none work (or poorly) try:
4. Freeze–thaw (without sonciation)
5. Freeze–thaw sonciation in the presence of detergent
6. Octylglucoside dialysis or cholate dilution
7. If nothing works, invent a new reconstitution procedure.

D. Multiple-Stage Reconstitutions

I have emphasized that proteins respond differently to detergents and that one reconstitution procedure may be suitable for one protein and not for another. Thus, a serious problem may arise in attempts to reconstitute a multicomponent system with different requirements for the individual components. We have used, in such cases, multiple-stage reconstitution. For example, for the reconstitution of bacteriorhodopsin with the mitochondrial or chloroplast proton pump for the generation of ATP, the most effective procedure is to first prepare bacteriorhodopsin vesicles by octylglucoside dilution (Racker *et al.*, 1979), followed by carefully controlled sedimentation, which allows collection of the proteoliposomes, leaving most of the free liposomes in the supernatant. This has the advantage that the second protein to be reconstituted has no choice; it must interact with the proteoliposomes. An additional useful feature is that the bacteriorhodopsin proteoliposomes are remarkably stable and can be kept frozen for many weeks. In the future there will be a greater need for multiple stage reconstitutions, e.g., when an ATP regenerating system is required (Racker, 1978) or pathways are assembled.

It should be apparent from this dull lecture that reconstitution of membranous proteins is largely empirical and not an intellectually stimulating excercise by itself. It is a means to an end, the end being a better understanding of how proteins function in membranes.

It is not always easy to convince students and postdoctoral fellows that the long road is the best. When we started on the reconstitution of oxidative phosphorylation, two of my postdoctoral fellows attempted to reconstitute the entire electron transport chain at once and did not listen to my advice to start with cytochrome oxidase.

Lesson 12:   How to stay at the bench

Don't force your postdoctoral fellows to listen to your brilliant ideas. Make your suggestions as persuasive as possible. I say to them: "These experiments are so exciting that if you do not want to perform them, I will do them myself." This policy has two advantages: (a) some postdoctoral fellows do get excited and try the experiments; (b) others do not, and have thus kept me busy in the laboratory for the past 30 years.

Lecture 3

What Can We Learn from Resolution and Reconstitution after the Natural Structure of the Membrane Has Been Destroyed?

> Traveller, there is no path;
> Paths are made by walking.
>
> **Antonio Machado**

I. What Are the Protein Components of the System and What Are Their Functions?

The physical separation of proteins of a functional complex is needed to demonstrate how many components are actually participating and to explore their specific roles. Genetic analyses have become increasingly important as an alternative approach to these two questions and will be discussed in detail in the lecture on the mitochondrial and bacterial proton pump. The combination of both approaches is showing great promise. The reconstitution of genetically defective, and thus partially resolved, membranes is likely to become the most efficient approach to develop assays for single protein components. It will avoid the pitfalls encountered using physical separations which often resolve more than one component. On the other hand, it is also likely that other pitfalls will be encountered using the genetic approach, such as aberrant dead-end assemblies or the presence of altered proteins, which may not be readily replaced by the normal component during reconstitution. Although such complications may make a mutant of this type unsuitable for reconstitution experiments, an analysis of the physical change in the mutant protein may yield important clues to its function.

It is therefore appropriate to stress that, as in the case of a detergent for solubilization and purification, the choice of a suitable mutant requires a horizontal survey until a mutant is found that works in a complementation assay.

The first example for the reconstitution of a multicomponent membrane complex is the proton pump of bovine heart mitochondria (Kagawa and Racker, 1971). Accidentally, it happens also to be the most complex ion transport system thus far encountered, since it requires at least 10 polypeptides in order to function in reconstituted proteoliposomes. There are peculiar pitfalls in the interpretation of such reconstitution experiments, since some of the components may be required for reconstitution (or *in vivo* for assembly) rather than for function. As judged by genetic analyses, the proton pump in microorganisms is somewhat less complex, consisting of only eight components (Futai and Kanasawa, 1980; Walker *et al.,* 1984). As I shall discuss in Lecture 5, these genetic analyses are also not without ambiguity. Cautious interpretations of multicomponent systems are required whether we use biochemical or genetic approaches. We are less likely to draw erroneous conclusions when we use both. There are features that become visible only when a complex structure is analyzed by multiple approaches.

Lesson 13:   How to analyze a three-dimensional structure

A sculpture such as the "Citizens of Calais" by Auguste Rodin (probably his masterpiece) can be viewed and admired from one side. But as we walk around, new and unexpected features appear. It is the mental reconstitution of these multiple views which allows us to perceive the concept of the whole, the "Gestalt" of the "Citizens of Calais."

I shall describe in a later lecture how the resolution and reconstitution of the mitochondrial proton pump was achieved and what we have learned from the analysis of the individual components. Now I would like to illustrate the need for the resolution of components by two examples that have not as yet been resolved.

The Na^+,K^+-ATPase of plasma membranes, as generally agreed upon, consists of an α and β subunit. The emergence of two different α subunits (α_1 and α_2) observed by SDS–PAGE analysis of brine shrimp Na$^+$,K$^+$-ATPase (Peterson *et al.*, 1982) may well have been caused by the use of a mixed cell population as a starting material for purification. Two different α subunits were found in different cells of rat brain (Sweadner, 1979) and adipocytes (Resh *et al.*, 1980), emphasizing the need for caution in invoking multiple forms of the same subunit in a complex. It is also generally agreed that the α subunit, seen as a band of about 100,000 daltons in SDS–PAGE, is the active site of interaction with both ATP and ouabain, a specific inhibitor of the Na$^+$,K$^+$-ATPase. There is virtually no information on the role of the β subunit. It appears to interact under some conditions with ouabain (Hall and Ruoho, 1980) but so does the elusive proteolipid present in preparations of Na$^+$,K$^+$-ATPase (Forbush *et al.*, 1978). The role of the β subunit in the plasma membrane pump is not likely to be elucidated without the physical or genetic resolution of the two subunits. While separation of native α and β subunits has thus far not been achieved, the translation of the α subunit of guinea pig ATPase by free ribosomes in cell-free systems has been reported (McDonough *et al.*, 1982). In view of the observation, to be discussed later, that the pure 100,000-dalton protein of the Ca^{2+}-ATPase of sarcoplasmic reticulum can function as a pump in reconstituted vesicles, it seems likely that the other associated proteins of the Na$^+$,K$^+$-ATPase, namely, the β subunit and the proteolipid, have secondary functions. In future reconstitution experiments the following possibilities for the function of these accessory proteins need to be considered: (a) a role in the organization within the membrane including a possible function in regulation, and (b) a role in the incorporation of the α subunit into the membrane. The β subunit could facilitate the proper folding required for entry of the α subunit into the membrane (Wickner, 1979), and thus serve as a receptor, such as required for the transport of mitochondrial proteins (Riezman *et al.*, 1983).

A similar unresolved problem exists in the case of the acetylcholine receptor. In spite of the successful reconstitution of this complex in several laboratories, the individual subunits have not been physically separated. No specific functions have been assigned to the four subunits, and similar considerations just discussed regarding the role of the β subunit of the Na$^+$,K$^+$-ATPase apply here. It also seems likely that the reconstitution of translation products of either cloned genes or of amplified cell-free protein synthesizing systems will be the most promising avenue of approach in unravelling the function of the subunits of the acetylcholine-activated channel (Anderson and Blobel, 1981; Mishina *et al.*, 1984).

II. What Are the Phospholipid Components and What Are Their Functions?

While there is general agreement on the high specificity of protein subunits that participate in various transport systems, there is relatively sparse information on a specific role of phospholipids in the function of membranous transport systems. Yet it is likely that the enormous complexity of the lipid composition of natural membranes is not accidental. Currently, reconstitution experiments are conducted at a very primitive level that may not shed light on the physiological role of specific phospholipids. We will have to learn to devise methods of phospholipid reconstitutions that will imitate not only the constitution but also the topology of protein–lipid interaction.

One of the primary purposes of performing reconstitution experiments is to establish the minimal requirements for function. Thus, a bacteriorhodopsin proton pump can function in a phospholipid bilayer of synthetic dimyristoyl phosphatidylcholine (Racker and Hinkle, 1974). The Ca^{2+} pump of sarcoplasmic reticulum can function in a liposome of either phosphatidylethanolamine (Racker, 1972b) or of phosphatidylcholine (Zimniak and Racker, 1978; Andersen *et al.*, 1983), depending on the method of reconstitution. The fact that the dependency on specific phospholipids varies with the method of reconstitution serves as a warning signal. I shall elaborate on this point below. When a method of reconstitution is used that requires the initial interaction with a detergent, there is almost always a detectable amount of residual detergent associated with the protein, thus rendering the system unphysiological. When sonication is used for reconstitution of proteins that are not delipidated, there are appreciable residual phospholipids present that are tightly associated with the native protein. In view of these considerations, the observations discussed later must be interpreted with caution.

The primary function of the phospholipids is to provide compartments that are impermeable to ions and other solutes that must be kept within the cell or organelle at appropriate concentrations, often at the expense of energy. The phospholipids must also serve as a suitable matrix for the embedding of proteins that facilitate the entry of desirable and the expulsion of undesirable solutes. The wide spectrum of different phospholipids and lipids suggests that there are additional functions, such as regulation of catalytic activities and of conformational changes, in the protein.

An analysis of the role of individual phospholipids has been carried out with the Ca^{2+} pump of sarcoplasmic reticulum (Knowles *et al.*, 1976). After stripping the purified Ca^{2+}-ATPase of most of its endogenous phospholipids with a detergent, the protein no longer catalyzed any of the reactions associated with its function. Phosphatidylcholine restored to the protein the ability to form phosphoenzyme in the presence of ATP and to hydrolyze ATP but was not sufficient for efficient Ca^{2+} transport. Phosphatidylethanolamine, on the other hand, was less effective in facilitating the first two functions but facilitated the formation of a suitable compartment and efficient Ca^{2+} transport. Both phospholipids used together during reconstitution gave the highest values for Ca^{2+} transport. The phosphatidylethanolamine used was isolated from a natural source and contained long-chain fatty acids. N-acetyl dilauryl phosphatidylethanolamine did not allow for the formation of a suitable compartment for Ca^{2+} transport and supported a very low ATPase activity, but it was the most effective phospholipid for the formation of phosphoenzyme. It therefore allowed the dissection of the enzyme-catalyzed reaction and the study of the first step, independent of subsequent events.

The chemical modification of the amino group of natural phosphatidylethanolamine was explored in experiments conducted in collaboration with H. G. Khorana (Knowles *et al.*, 1975). The acetylation of phosphatidylethanolamine did not impair the ability of the phospholipid to form tight vesicles capable, e.g., of proton pumping with bacteriorhodopsin. However, the Ca^{2+} transport activity of reconstituted vesicles was lost completely but could be restored by further incorporation of a hydrophobic alkylamine such as stearylamine or oleoylamine. Even the ATPase activity was not well supported by the N-acetylphosphatidylethanolamine unless an alkylamine was incorporated as well. Dicetylphosphate inhibited this system and could be quantitatively titrated against stearylamine, suggesting a role of charges on the assembly or function of the pump. Similar observations were recorded with the mitochondrial proton pump, which functioned only with N-acetylphosphatidylethanolamine when an alkylamine was present. We could show in this case that the N-acetylphosphatidylethanolamine contributed to the assembly or activity of the pump, since phosphatidylcholine, which was capable of forming a compartment, did not function even in the presence of stearylamine unless acetylphosphatidylethanolamine was present as well. These studies show how reconstitution experiments can shed light on the contribution of individual phospholipids and allow us to explore specific chemical groups within the phospholipids. However, we know that a Ca^{2+} pump can function with phosphatidylcholine as the only phospholipid when the vesicles are prepared by the freeze–thaw sonication procedure. It is therefore conceivable that the multitude of phospholipids may be more important for either insertion of the proteins into the biological membrane or for regulation and efficiency of pump operation rather than for the basic physiological function. On the other hand, we have to consider the possibility that residual phospholipids attached to the protein obscure the true phospholipid requirements when a sonication procedure is used.

Of particular interest are reconstitution experiments with endogenous phospholipids. Bacteriorhodopsin reconstituted with $H.$ $halobium$ phospholipids at an acid pH yielded right side out vesicles that pumped protons out (Happe et $al.$, 1977). Remarkably stable bacteriorhodopsin proteoliposomes were prepared with endogenous polar phospholipids of $H.$ $halobium$ (Lind et $al.$, 1981). Moreover, they exhibited an ion selectivity at high salt concentration giving rise to much higher rates and extent of H^+ translocation with NaCl than with KCl. Also greater stimulation of H^+ transport by valinomycin was seen than with vesicles reconstituted with soybean phospholipids. In several other respects, the vesicles reconstituted with endogenous lipids appear to resemble the properties of bacteriorhodopsin in the native membranes more closely than when heterologous lipids were used.

Similar differences were noted in the reconstitution of the lactose transporter from $E.$ $coli$, which appears to prefer its own lipids to that of soybeans (Newman and Wilson, 1980).

Lesson 14: How to reconstitute more physiologically

It is indeed remarkable how many membrane proteins are willing to become associated with crude soybean phospholipids. Because of the bulk availability and low price of these phospholipids, they have become a widely used source for membrane reconstitutions. We should not take promiscuity for granted, since some membrane proteins are more selective. The closer we want to imitate the physiological properties of a membranous system, the more attention we need to pay to its natural environment.

III. What Is the Role of Asymmetry? How Do We Achieve It in Reconstitution? How Do We Measure It?

It is obvious that, in the case of ATP-driven pumps of the plasma membrane and sarcoplasmic reticulum, the organization of the integral membrane protein must be asymmetric with the ATP site facing the cytoplasm. In the case of the mitochondria, the ATP site faces the matrix; therefore, a special transport system for ADP and ATP is provided in order to allow communication with the cytoplasm. With ion transporters catalyzing ATP/ADP or channels such as the tetrodotoxin-sensitive Na^+ channel, the need for asymmetry is not quite as obvious because of prevailing concentration gradients, but it appears in all cases that have been examined that such transporters are also asymmetric. This can be established by demonstrating site specificity in response to water soluble inhibitors. Atractyloside inhibits the ADP/ATP transport of mitochondria but not inside out submitochondrial particles when added from the outside (Shertzer and Racker, 1974). N-ethylmaleimide, which does not traverse the mitochondrial inner membrane, inhibits P_i transport only from the cytoplasmic side (Rhodin and Racker, 1974), whereas hydrophobic maleimides inhibit from either side of the membrane. Tetrodotoxin inhibits the Na^+ channel only from the outside (Villegas et al., 1977). These observations not only demonstrated asymmetry but allowed for the design of an assay to quantitate the extent of asymmetry achieved in reconstitutions. Why are these transport systems which often operate in either direction organized asymmetrically? A good reason for the asymmetry arises from the fact that channels can be controlled from the outside (e.g., by an agonist) and from the inside (e.g., by phosphorylation) by different mechanisms. Thus asymmetry induced by the presence of specific structures on one side and not the other can serve the same purpose as allosteric control mechanisms of soluble enzymes. It can also, as in the case of allosteric enzymes, help to control the flux by selective interaction with natural regulators. The glycosylation processes that take place in the ER and Golgi membranes contribute to the asymmetry of membrane proteins. The protruding carbohydrates moreover help to protect plasma membrane components against a hostile environment, as in the case of bacterial polysaccharides.

How do we achieve asymmetry in reconstitutions? In the case of the reconstitution of oxidative phosphorylation this was a serious problem that had to be solved, as will be described in Lecture 5. Fortunately, in the case of most reconstituted systems, asymmetry or predominant asymmetry is achieved by the very nature of the hydrophobic membrane proteins. Thus it is usually the reconstituted protein and not the reconstitutor who makes the decision of directionality. In the case of the mitochondrial ATPase and the plasma membrane Na^+,K^+-ATPase, the reconstituted vesicles are mainly inside out; in the case of the Ca^{2+} pump of sarcoplasmic reticulum they are right side out. Is this a coincidence or is there a message here? The common denominator is that, in all known cases of reconstituted ATP-driven pumps, the ATP interacting site in the majority of vesicles faces the outside. Is it not possible and indeed likely that the other side, containing the ion channel that is embedded in the membrane, is more hydrophobic and therefore is preferentially incorporated into the lipid bilayer? This is particularly so in the case of F_1 or CF_1, which are very hydrophilic proteins that protrude from the membrane into the aqueous environment. What about bacteriorhodopsin? Why is it primarily oriented in the inside out configuration in reconstituted proteoliposomes? Could we attach to it a hydrophobic polypeptide at the internal end that would favor its incorporation in the right side out configuration, as if directed by a signal polypeptide? Or could we alter the external end of the protein, rendering it more hydrophilic, e.g., by attaching an antibody, and thus reducing the probability of it entering first during reconstitution? We have initiated experiments to probe these possibilities. Is it likely that the natural insertion is aided by a signal peptide that is cleaved by an external protease?

IV. How Do We Measure the Extent of Scrambling during Reconstitution?

Once again no general guidelines can be enumerated since in each system advantage should be taken of its biological properties. The simplest assay can be designed when both impermeable and permeable inhibitors are available. In the case of the adenine nucleotide transporter, submitochondrial particles, which are inside out, are resistant to the hydrophilic atractyloside but sensitive to the hydrophobic bongkrekic acid. Reconstituted vesicles, which are right side out, are sensitive to both inhibitors (Shertzer and Racker, 1976). In the case of the P_i transporter of submitochondrial particles, the ion uptake is resistant to the water-soluble N-ethylmaleimide but sensitive to hydrophobic maleimides, thus confirming the inside out configuration of the particles (Rhodin and Racker, 1974). What can we use when such convenient inhibitors are not available? In the case of ATP-driven pumps that form covalent intermediates, one can determine with [^{32}P-γ]ATP the extent of phosphorylation before and after reconstitution. It is essential in such experiments to remove any unincorporated ATPase. The presence of loosely bound ATPase would greatly reduce the sensitivity of the assay, which depends on detecting a difference between two large numbers. Thus, 10 or 15% scrambling could easily be missed. We have successfully applied this method to the Ca^{2+}-ATPase of sarcoplasmic reticulum, which reconstitutes efficiently, and removal of unreconstituted ATPase by sucrose gradient centrifugation is feasible (Knowles and Racker, 1975a). It has thus far proved more difficult in the case of the Na^+,K^+-ATPase, which was not as efficiently incorporated, and removal of ouabain-sensitive ATPase (a measure for the unincorporated enzyme) was more difficult. In this case (as in the case of the mitochondrial proton pump), it would be of considerable advantage to have a functional assay available to measure the extent of asymmetry. For this very purpose we initiated a number of years ago the isolation and reconstitution of the adenine nucleotide transporter. This could allow us to set up an ATP-regenerating system in proteoliposomes using, for example, phosphoenolpyruvate and pyruvate kinase externally in the presence of catalytic amounts of ADP, both outside and inside of the reconstituted vesicles. Alternative approaches will be discussed in Lecture 6. In the case of the mitochondrial ATPase, the coreconstituted nucleotide transporter allowed us to measure the extent of right side out configuration by determining the amount of atractyloside-sensitive cleavage of ATP (Banerjee and Racker, 1979).

A simple functional assay of scrambling is available in the case of cytochrome oxidase. Measurements of oxygen uptake on addition of impermeable cytochrome c allows for a quantitative evaluation. If on addition of valinomycin and nigericin respiration was maximal, i.e., the same as before reconstitution or the same as after addition of a detergent such as peroxide-free Tween 80 or Emasol that open the liposomes to cytochrome c without inhibiting oxidase activity or causing autooxidation of ascorbic acid, we concluded that reconstitution is in the right side out configuration. Indeed, this was found to be the case when either the incorporation procedure or cholate dialysis was used for reconstitution. However, when oxidized cytochrome c (but not reduced cytochrome c) was present during cholate dialysis, the reconstituted vesicles were scrambled with about 40% in the inside out configuration (Carroll and Racker, 1977). A kinetic analysis of this phenomenon showed that addition of cytochrome c can be delayed for several hours after initiation of dialysis without altering the results. This suggests that the cholate dialysis procedure may in fact be a variant of the incorporation procedure and may require the formation of perhaps metastable liposomes prior to the entry of the protein. This effect of cytochrome c facilitating scrambling turned out to be a crucial feature in the reconstitution of oxidative phosphorylation, which would not have been possible without some of the cytochrome oxidase oriented in the inside out configuration. Establishing the degree of scrambling in other systems is often difficult. In the case of reconstituted vesicles catalyzing site I of oxidative phosphorylation, spectral changes induced by impermeant reagents were used to determine the degree of scrambling. Here, as in the case of cytochrome oxidase, the degree of scrambling was influenced by the presence of other components, such as the proteolipid of the inner mitochondrial membrane (Ragan and Racker, 1973). The use of impermeant probes to evaluate the asymmetry of reconstituted vesicles includes the use of chemicals such as p-diazonium benzoate (Eytan and Broza, 1978) or a radioimmune assay as in the case of HLA antigens (Curman et al., 1980).

A most important approach to this problem has been taken by Khorana and his collaborators (Lind et al., 1981) using proteolytic digestion. In these studies, which may serve as a model for future experiments on reconstitution, bacteriorhodopsin was exposed to chymotrypsin before and after reconstitution. Current models of bacteriorhodopsin structure within the membrane predict that the cleavage site should be on the inside of reconstituted vesicles that are inside out. Only little cleavage was observed unless the vesicles

TABLE 3-1 Orientation of Membrane Proteins in Reconstituted Vesicles

| | Oxidative phosphorylation site 1 | | Cytochrome oxidase | | Ca^{2+} pump | Bacterio-rhodopsin proton pump | ADP–ATP transporter | |
| | | | | | | | SMP[a] | Proteo-lipo-somes |
	+HP[a]	−HP	+Cyto[a] c	−Cyto c				
Right side out	50	80	60	100	100	0	0	100
Inside out	50	20	40	0	0	100	100	0

[a] HP, hydrophobic protein, crude preparation of F_0 from bovine heart mitochondria; Cyto, cytochrome; SMP, submitochondrial particles.

were opened with Triton X-100 confirming the inside out configuration of these vesicles.

A partial list of orientations in reconstituted proteoliposomes is shown in Table 3-1.

How can we use reconstitutions to learn more about the nature of the transport system and its mechanism of action? We shall deal with this problem in the next lecture together with a discussion of pitfalls encountered in reconstitution experiments.

Lecture 4

Analyses of Reconstituted Vesicles: Pitfalls and Obstacles

Experience is the name
everyone gives to his mistakes.

Oscar Wilde

I. Analyses of Reconstituted Vesicles

A. Mechanisms of Respiratory Control

I mentioned in the last lecture that addition of nigericin and valinomycin to reconstituted cytochrome oxidase vesicles elicits maximal respiration. What do we learn from these observations? When we first analyzed these proteoliposomes (Hinkle *et al.,* 1972) we noticed a marked loss of cytochrome oxidase activity, in comparison with the nonreconstituted enzyme, as measured by the oxidation of ascorbate in the presence of cytochrome *c*. On addition of a proton ionophore, such as compound 1799, the respiratory rate increased greatly. Thus, the reconstituted vesicles exhibited respiratory control. This observation allowed us to examine the mechanism of this phenomenon by selectively collapsing the membrane potential and the Δ pH, the two components postulated by Mitchell (1966) to control respiration. Valinomycin in the presence of K^+ collapses the membrane potential; nigericin catalyzes a K^+/H^+ exchange and collapses the Δ pH (Pressman, 1976). When either valinomycin or nigericin was added to the reconstituted cytochrome oxidase vesicles, only a marginal stimulation of oxygen uptake was noted (Table 4-1). When both were added together, a 6- to 12-fold stimulation of respiration was recorded in optimally reconstituted cytochrome oxi-

TABLE 4-1 **Respiratory Control in Cytochrome Oxidase Vesicles**[a]

	ng atom oxygen/min	Respiratory control ratio
Cytochrome oxidase vesicles	28	—
+ 0.2 μg Valinomycin	30	1.07
+ 0.2 μg Nigericin	54	2.0
+ 40 μg Oleic acid	61	2.1
+ Valinomycin + nigericin	205	7.3
+ Valinomycin + oleic acid	188	6.7

dase vesicles. As can be seen from Table 4-1, oleic acid had an effect similar to that of nigericin, requiring valinomycin for maximal effectiveness. This may have significant physiological implications with respect to fatty acids as potential natural uncouplers and mediators of heat production in thermogenesis, e.g., during hibernation. The contribution of fatty acids and the 32,000-dalton "uncoupler" protein (Nicholls, 1983) needs to be elucidated by reconstitution experiments.

B. Cytochrome Oxidase Vesicles as Proton Pumps

There has never been a period of extended peace in the field of oxidative phosphorylation. One of the current battles concerns the presence of a proton pump in reconstituted cytochrome oxidase vesicles.

Some years ago we were intrigued by the possibility that cytochrome oxidase functions as a proton pump. This idea was stimulated by the observation (Carroll and Racker, 1977) that vesicles reconstituted with cytochrome oxidase purified by ammonium sulfate precipitation in the presence of cholate and Tween 80 showed a marked inhibition of oxygen uptake by DCCD, which was reversed by uncouplers. The excitement about this "evidence" for the existence of a proton pump analogous to F_1F_0 soon evaporated when we found that the DCCD inhibition was due to a reversal of a stimulation caused by Tween 80.

The studies of Wikström (1977), Wikström et al. (1981a,b), and Lehninger et al. (1979) pointing to the presence of proton pumps in the respiratory chain called for a reexamination of the mechanism of cytochrome oxidase action, as formulated by the chemiosmotic hypothesis (Mitchell, 1966). Analyses of cytochrome oxidase vesicles by Wikström and Saari (1977), Sigel and Carafoli (1980), Coin and Hinkle (1979), Casey et al. (1979), and Saraste et al. (1981) represent probably the strongest evidence that cytochrome oxidase functions as a proton pump rather than as a simple redox loop. Indeed, the question of whether subunit III of cytochrome oxidase is the key to the proton pump mechanism should be solved in the near future by reconstitution experiments. This problem and the reconstitution of bacterial cytochrome oxidase will be discussed in a later lecture.

C. Channel versus Mobile Carrier Mechanisms

Bacteriorhodopsin reconstituted with synthetic phospholipids of known transition temperatures was analyzed for its capacity to pump protons at different temperatures (Racker and Hinkle, 1974). As shown in Table 4-2, proton pumping in bacteriorhodopsin proteoliposomes was markedly depressed by nigericin when measured at 30°C above the transition temperature of the synthetic dimyristoyl phosphatidylcholine used for reconstitution. Nigericin, which is a mobile carrier, was, however, ineffective when tested at 5°C, at which temperature the membrane is frozen. In contrast, gramicidin functioned as a channel equally well below the transition temperature, provided it was added to the phospholipid above the transition temperature. The response of bacteriorhodopsin to temperature was similar to that of gramicidin. It formed a channel which functioned at 5°C. In fact, in the presence of nigericin, bacteriorhodopsin functioned as a proton pump only below the transition temperature of the phospholipid, thus representing an artificially created paradox—a pump that only functions effectively in the cold.

TABLE 4-2 Bacteriorhodopsin, a Channel
Pump

Dimyristoyl phosphatidyl-choline vesicles	H^+ pump activity at	
	30°C	5°C
Control	100%	105%
+ gramicidin	18%	20%
+ nigericin	16%	105%

D. Electrogenic Ion Movements

Vesicles reconstituted with the Ca^{2+}-ATPase of sarcoplasmic reticulum (Racker, 1972b) are not representative of the physiological process. They did not transport Ca^{2+} significantly unless preloaded with oxalate or phosphate, which serve as Ca^{2+} traps. The reconstituted vesicles did not, like native sarcoplasmic reticulum vesicles, respond to externally added oxalate or phosphate. Their "unnatural" properties led us to propose that we had destroyed the natural pathway of a cotransport of Ca^{2+} and P_i. They provided us with an assay for the reconstitution of the P_i transporter (Carley and Racker, 1982) by measuring ATP-dependent uptake of $^{32}P_i$. We have recently encountered a very analogous situation in the case of the H^+ pump of clathrin-coated vesicles (Xie et al., 1983), which is electrogenic and functions only if chloride ions move together with the protons. Thus, an ATP-dependent chloride uptake can be experimentally demonstrated. It appeared from the experiments with purified Ca^{2+}-ATPase from sarcoplasmic reticulum that the reconstituted vesicles were deprived of ion flux capacities present in the native vesicles. This allowed us to establish unambiguously that this pump is electrogenic (Zimniak and Racker, 1978). By varying the ion concentration during reconstitution and assay, different K^+ gradients across the membrane were established, permitting the creation of controlled membrane potentials in the presence of valinomycin. The membrane potential, which had no effect on the rate of Ca^{2+} transport (about 60 mvolt positive inside the vesicles), was taken as a reflection of the membrane potential generated by the pump during Ca^{2+} translocation. This value was in agreement with other measurements, e.g., using a fluorescent probe.

Another example for the operation of an electrogenic ion transport is the adenine nucleotide transporter of mitochondria, which was reconstituted into proteoliposomes. The response to valinomycin in the presence of K^+ was used to show the electrogenic nature of this transporter in reconstituted vesicles (Shertzer and Racker, 1976). The possible pitfalls of this approach will be discussed later.

E. The ATP-Driven Proton Pumps

Mitochondria and chloroplasts are complex structures with various ions moving in different directions. To avoid some of the complexities, Dewey and Hammes (1981) have performed kinetic studies on the chloroplast proton pump in vesicles reconstituted together with bacteriorhodopsin. This permitted them to establish a constant transmembranous electrochemical proton gradient by illumination. Among other observations it was found that the Michaelis constants for ATP, ADP, and P_i were independent of the Δ pH, whereas maximum velocities strongly depended on it. ATP synthesis was enhanced when ATPase activity was suppressed by the pH gradient. The H^+/ATP ratio was 3, confirming previous observations with intact chloroplasts (Portis and McCarty, 1974). The most interesting conclusion emerging from these kinetic studies pertaining to the mechanism of action of the proton pump is that the three protons are not pumped in a single step. An attractive model was proposed that involves conformational changes in the protein induced by substrate binding that alter the environment of ionizable groups on the outside and inside, respectively, resulting in the uptake and release of protons. Another important conclusion was that the vesicles reconstituted with crude soybean phospholipids were very leaky to protons. Vesicles prepared with purified lipids were much less permeable to protons and showed marked stimulation of H^+ pumping by valinomycin (Lind et al., 1981).

II. Pitfalls and Recommended Cautions

A. Residual Activity of Native Vesicles

What we cannot learn from reconstituted systems is how the transporter really operates in vivo. They have, as we have just pointed out, some advantages because (a) they allow us to analyze some of the properties of the transporter that are difficult to analyze in native membranes, and (b) they may provide us with a specific assay that allows us to conclude that we have indeed preserved functional activity. One of the most frequent pitfalls in reconstitution experiments starting with native membrane fragments is the often remarkable resistance of membrane vesicles to disruptive forces. For example, I have observed light-dependent proton movements in particles from H. halobium mutants (that lack bacteriorhodopsin but contain halorhodopsin) (Schobert and Lanyi, 1982; MacDonald et al., 1979) even after they were exposed to freeze–thawing and prolonged sonication. Since they remained active when sonicated, either in the presence or absence of added phospholipids, it was apparent that these vesicles were remarkably resistant to such vigorous treatment. Yet the same vesicles were destroyed at extremely low concentrations of detergents, e.g., minute amounts of lysolecithin. On the other hand, we have observed ATP-dependent proton movements in clathrin-coated vesicles (Xie *et al.*, 1983) that had been exposed to more than 1% octylglucoside or 1% sodium cholate. Because of this unpredictability of response, it is advisable to search for "unnatural properties" in reconstituted vesicles that assure us that we have indeed achieved reconstitution and are not dealing with residual native vesicles. Particularly useful in this respect are water-soluble inhibitors which act only on one side of the membrane, such as ouabain, atractyloside, or tetrodotoxin, which also serve as tools for the quantitative assay of degree of scrambling.

Lesson 15:   On the individuality of native membrane vesicles

The differences between native membrane vesicles from different sources are truly remarkable. Some vesicles that are disrupted by minute amounts of lysolecithin are deaf to sonic oscillation. Other vesicles go to pieces when exposed to sonication but resist high concentrations of detergent (e.g., 1% cholate). What is meat for one vesicle may be poison for another. Since liposomes without protein become leaky to ions in the presence of 1% cholate, we must conclude that either stability is conferred by specific protein–phospholipid interactions or that some proteins facilitate resealing.

When the reconstituted vesicles have the same orientation as the native vesicles, it may be difficult to tell them apart without physical separation. If "reconstitution" takes place without added phospholipid, start all over again.

B. Reconstitution of Nicked or Derivatized Proteins

One of the obstacles in the life of biochemists is that cell-free extracts contain proteolytic enzymes. It is therefore difficult to be certain that after tedious purifications the proteins used in reconstitution experiments are truly native. A dramatic example of the pitfalls encountered is the story of nicked acetylcholine receptor. This receptor was purified from Torpedo membranes and shown to have four subunits according to SDS–PAGE analysis: α (40,000 daltons), β (48,000 daltons), γ (58,000 daltons), and δ (64,000 daltons), with a stoichiometry of $\alpha_2\beta\gamma\delta$ for the monomer with a molecular weight of 250,000 (Reynolds and Karlin, 1978). It was subsequently reported that a preparation consisting mainly of the 40,000-dalton subunit according to SDS–PAGE analysis was active in reconstituted vesicles, suggesting that the other subunits may not be essential for function (Changeux *et al.*, 1979). Similarly, a receptor preparation missing the γ subunit was found to be reconstitutively active (Huganir *et al.*, 1979). These observations were shown to be caused by nicking of the receptor protein by proteases (Huganir and Racker, 1980). Indeed, a pronase-treated preparation, which contained only 23,000- and 10,000-dalton polypeptides according to SDS–PAGE analysis, was still reconstitutively active and exhibited the characteristic property of desensitization. It was shown that treatment with pronase, though giving rise to extensive nicking of all polypeptide subunits, did not dissociate the receptor complex, as was shown by sucrose gradient centrifugation. The personality defects of the complex were only detected by SDS–PAGE analysis. It is therefore important to remember that we need to be cautious in the interpretation of data obtained by SDS–PAGE analysis.

Lesson 16: Personality defects in proteins

Watch out for personality defects in proteins used for reconstitutions. Protease-sensitive proteins must be shielded against injury by a multitude of natural and synthetic protease inhibitors. A nicked protein may appear deceptively normal even in reconstitution experiments, whereas psychoanalysis by SDS–PAGE electrophoresis reveals a shattered personality. Treatment of such defective proteins has proven to be about as effective as treatment of personality defects in man. Prevention is better than cure.

C. Other Personality Defects

Another example is the Ca^{2+}/Na^+ antiporter from plasma membranes of bovine heart which was reconstituted after extraction with cholate and partial purification (Miyamoto and Racker, 1980). Prior to these successful experiments, we had achieved reconstitution of a

Ca^{2+}/Na^+ exchange following extraction of plasma membranes with Triton X-100. We decided to reject these experiments as "unphysiological", because the reconstituted proteoliposomes failed to exhibit the phenomenon of Na^+ control (Reeves and Sutko, 1979). An important advance in the purification of the Ca^{2+}/Na^+ antiporter was achieved recently (Wakabayashi and Goshima, 1982). Treatment of vesicles with pronase resulted in a spectacular purification of Ca^{2+}/Na^+ transport of about 30-fold. Although it is quite possible that the transporter is more resistant to pronase than 97% of all other proteins in these preparations, there is an obvious possibility for serious personality defects in view of the experiments on acetylcholine receptor described above.

As pointed out in the second lecture, the choice of detergent is important, since some proteins may be altered in very subtle ways that are not always detectable by enzymatic assays. One example is cytochrome oxidase extracted and purified in the presence of deoxycholate. In spite of its high enzymatic activity, such an enzyme is reconstitutively inactive. This personality defect actually appears to be curable, since I have been able to recover reconstitutively active preparations after displacing deoxycholate by refractionation of the enzyme with ammonium sulfate in the presence of cholate (E. Racker, unpublished observations, 1975). A thoroughly analyzed personality defect is the case of succinate dehydrogenase. The enzyme was shown to be reconstitutively active when isolated in the presence but not in the absence of succinate (King, 1963). In elegant studies, Baginsky and Hatefi (1969) traced this personality defect to a loss of nonheme iron which could be replaced by appropriate biochemical manipulations. Thus, loss of reconstitutive activity is a pathology of protein structure worthy of investigation, since it gives us information on the allotopic properties and on other functionally important features of the protein.

Personality defects have been induced in ATP-driven proton pumps of mitochondria and chloroplasts by either mild proteolytic digestion or exposure to chemicals. F_1 exposed to mild tryptic digestion still combines with the mitochondrial membrane giving rise to an oligomycin-sensitive ATPase, but it no longer functions as a coupling factor capable of catalyzing ATP generation (Horstman and Racker, 1970), presumably because of an internal proton leak in F_1. A similar leak has been introduced into chloroplast CF_1 by bifunctional maleimides (McCarty, 1979).

Purification of CF_1 released from chloroplasts by chloroform treatment yielded a preparation that failed to restore photophosphorylation when added to thylakoids that were partially deficient in CF_1 (Nelson and Karney, 1976; Younis *et al.*, 1977), although binding of CF_1 to the membrane still occurred (Andreo *et al.*, 1982). Photophosphorylation was restored by addition of the δ subunit of CF_1. Thus, a personality defect by subunit deficiency can be introduced by mishandling a protein during purification. As pointed out earlier, such defects can yield important clues to the role of the missing or defective subunit.

Lesson 17: Purity

Like beauty, purity of a protein is in the eye of the beholder and varies with the seasons. The day before yesterday a single peak in the analytical ultracentrifuge was a standard of purity; yesterday it was a single band in SDS–PAGE stained with Coomassie blue; today it has to be pure by silver staining; tomorrow it may have to be as pure as gold.* Try to add a potent protein kinase with hot [γ-^{32}P]ATP to your protein. You may find some bands which "Hi-Ho, Silver" has missed. Tomorrow someone will find a dissociating reagent that is better than SDS, and some of the "pure protein" may turn out to be a couple of hydrophobic friends.

D. Reconstitutions of Unknown Functions

It sometimes happens in the course of research that one looks for one thing and finds another. We have reconstituted two proteins that are waiting for the assignment of a function. Such examples should probably not be considered as pitfalls unless the investigator fails to see that he is not getting what he was looking for. The first example is the isolation and reconstitution of a P_i transporter from bovine heart mitochondria (Banerjee and Racker, 1979). After reconstitution into liposomes, the partially purified protein catalyzed an active P_i–OH exchange which was sensitive to SH reagents, such as *N*-ethylmaleimide. The marked stimulation by valinomycin suggested that the catalyzed ion exchange is electrogenic, which was contrary

---

* I was informed by Dr. G. Schatz after "completion of this manuscript" that such a method exists: protein A with bound colloidal gold—as an immune blot test.

to the rather persuasive evidence that in intact mitochondria it is not (Chappell, 1968; Mitchell and Moyle, 1969). There are two possible explanations for this discrepancy. The first one, rejected by the establishment, is that the persuasive evidence for the electroneutrality is an experimental artifact and that there is a biological coupling between P_i and ADP entry which has escaped detection. Such an arrangement would be attractive and economical because it would eliminate the energy expenditure of ADP translocation and allow the entry of ADP and P_i to be electrically balanced by the export of ATP. The second explanation, which is acceptable to the establishment, is that we are dealing either with an artifact of reconstitution or with an unknown function, perhaps involving an electrogenic export of P_i from mitochondria. In any case, the availability of the reconstitutively active and electrogenic P_i transporter has been used in our laboratory as a tool to demonstrate the reconstitution of an atractyloside-sensitive ATPase in the right side out configuration (Banerjee and Racker, 1979).

*Lesson 18: Early decisions**

> If you start on a new project and do not find the activity you want, study the activity you get; it is probably more interesting than what you had expected.

A second example of an unexpected reconstitutively active protein emerged during attempts to isolate the electrogenic carrier of Ca^{2+} transport from mitochondria (Saltzgaber-Müller et al., 1980). Using reconstituted cytochrome oxidase vesicles as a test assay, we discovered that they catalyzed a Ca^{2+} uptake that was dependent on addition of ascorbate and cytochrome c. Treatment of cytochrome oxidase with chymotrypsin did not impair its catalytic activity, the phenomenon of respiratory control, or its proton pump activity after reconstitution, but greatly reduced Ca^{2+} uptake by the vesicles. Chymotrypsin destroyed (among others) a polypeptide with an M_r of 8800, which was extracted with 60% ethanol, and restored Ca^{2+} uptake activity when added to the altered vesicles. However, several observations were inconsistent with the identification of this protein with the electrogenic Ca^{2+} carrier. Valinomycin in the presence of K^+ failed to inhibit Ca^{2+} transport; in fact, it stimulated it. Secondly, ruthenium red did not inhibit Ca^{2+} uptake in contrast to

---

* Contributed by Dr. C. Miller.

its effect in mitochondria (Moore, 1971). After eliminating several other explanations for these observations, we must conclude that we have reconstituted an unknown function or that we are dealing with an artifact of reconstitution.

E. The Deceptive Roles of Phospholipids

Perhaps the most attractive trap to fall into is to draw conclusions from reconstitution experiments on the role of phospholipids and proteolipids, as discussed earlier. What needs to be reemphasized is the need to dissociate effects on reconstitution from effects on the biological process per se. This is easier said than done, and more attempts are needed to alter the properties of phospholipids *in situ*. A serious inherent difficulty is caused by the tight association of endogenous phospholipids with highly purified membranous proteins, as in the case of cytochrome oxidase and the Ca^{2+}-ATPase of sarcoplasmic reticulum. Thus, more extensive work is needed with completely naked proteins using a variety of different reconstitution procedures before a definitive answer with respect to phospholipid specificity can be given.

F. Pitfalls Encountered in Assays of Reconstituted Vesicles

There are numerous methods for the separation of native membrane vesicles from solutes, but some of them are not suitable for the separation of proteoliposomes from solutes. Even more treacherous is the fact that suitability of separation methods varies with the composition and size of the proteoliposomes and even with the solute under investigation. Thus, in each case it is necessary to establish that the assay yields quantitatively accurate data and is proportional to protein concentration.

Basically there are three methods widely used for assay of reconstituted vesicles: (1) Filtration, e.g., through Millipore filters; (2) separation by ion exchange column; and (3) separation by molecular sieves.

1. Filtration through Millipore Filters

This method is suitable for ions and is particularly useful with uncharged solutes such as glucose. The method has been successfully used for the uptake of Ca^{2+} into phosphate- or oxalate-loaded reconstituted vesicles (Racker, 1972b). A comparison of this method with method 2 or 3 yielded virtually identical rates of Ca^{2+} uptake (Gasko et al., 1976). Millipore filtration has also been successfully used for measurements of glucose uptake (Kasahara and Hinkle, 1976). It has the added advantage of allowing one to concentrate vesicles when a detergent dilution method is used, as in the case of lactose transport reconstitution (Newman et al., 1981). Yet in many instances the method is unreliable because a large fraction of small proteoliposomes are not retained by Millipore filters (0.45 or 0.22 μm). On the other hand, with giant liposomes the rate of filtration through Millipore filters may be too slow for accurate results, and a search for other filters is indicated. Particularly confusing were unpublished observations in our laboratory which showed that the Millipore filtration was satisfactory with one batch of phosphatidyl-ethanolamine and not with another, although both were "pure" according to thin-layer chromatography. Small amounts of undetected impurities may have influenced the size or state of aggregation of reconstituted proteoliposomes. Moreover, suitability varies with the reconstitution method used and particularly with the size of reconstituted proteoliposomes. Thus, the large proteoliposomes obtained by the freeze–thaw or the freeze–thaw sonication procedures are retained by Millipore filters. If the fraction of small proteoliposomes that may filter through is either negligible or not functional in transport, the method is valid. We have explored a variety of methods that lead to aggregation of small proteoliposomes, such as treatment with $HgCl_2$ or protamine prior to filtration, procedures used in other laboratories for the filtration of proteoliposomes. We have invariably encountered lower values of transport activities (mainly caused by induced leakage) when we compared this procedure with other methods of analysis, such as gel filtration or ion exchange columns.

2. Separation by Ion Exchange Columns

This method (Gasko et al., 1976) is recommended as the method of choice when movements of cations are measured, provided the recommended precautions are taken to prevent retention of liposomes. It is also the method of choice for measurements of anion fluxes, such as chloride, phosphate, or thiocyanate. However, difficulties have been encountered in our laboratory with measurements of lactate transport because of incomplete retention of radioactivity on the column, presumably because of the presence of uncharged impurities. The use of mixed-bed resins did not solve the problem. Similar difficulties have been encountered with radioactive amino acids (M. Kilberg, personal communication).

3. Separation by Molecular Sieves

Separation by passage through Sephadex columns or other molecular sieves, in spite of some drawbacks, is often the method of choice in the case of uncharged substrates. Our experience with measurements of lactate transport have revealed another pitfall encountered with proteoliposomes, namely leakage of the solute even in the absence of protein or in the presence of nonspecific hydrophobic protein. It is therefore essential to standardize the Sephadex separation method with preloaded liposomes to establish that they retain the solute. Moreover, when leakage problems of this type emerge it becomes absolutely essential to use a specific transport inhibitor which does not induce leakage in liposomes. Mercurials, which have been successfully used as inhibitors of lactate transport in intact cells (Spencer and Lehninger, 1976), induce leakage of lactate from liposomes (M. Newman, personal communication).

The obstacles we have encountered in attempts to reconstitute the lactate transporter of plasma membranes of Ehrlich ascites tumor cells have induced us recently to attempt the reconstitution of various types of giant proteoliposomes. Since these attempts are still in a stage of infancy (or fancy), we postpone their discussion to the last lecture, projecting glimpses into the future of reconstitution. It is apparent from all that has been said and done with reconstitution of proteoliposomes that there is still a long and winding road ahead of us.

What can we learn from resolution and reconstitution after the natural structure of the membrane has been destroyed?

1. How many protein components are essential for the basic function of the system?

2. What are their functions?

3. What is the role of other protein components found in association with the purified complex?

4. What are the minimal phospholipid requirements? Are there specific phospholipids in association with the protein that are essential or beneficial for function?

5. How do variations in phospholipid composition affect the efficiency of the process?

6. Are we dealing with an electrogenic system with a carrier, a channel, or a complex mixture of both?

7. What are the kinetic parameters of the reconstituted system when measured without perturbations by other processes?

Lecture 5

The ATP Synthetase of Oxidative Phosphorylation

It is the customary fate
of new truths to begin as
heresies and to end as
superstitions.

Thomas Henry Huxley

Twenty years ago the field of oxidative phosphorylation was in a state of confusion—not just facilitated confusion but active confusion. Hypotheses were dragged up the hill, which was surrounded by a fog of rhetoric and equations. Anyone who was not thoroughly confused just did not understand the situation. No one agreed with anyone; what happened in Wisconsin did not happen in Baltimore or in Los Angeles, and vice versa. When I lectured at that time, I entitled my seminar "Oxidative Phosphorylation in New York City." The field was characterized, in the words of Jeff Schatz, by weak data and strong characters.

In the midst of this, Mitchell appeared with the proposal of the chemiosmotic hypothesis based on no facts. I ignored it. It was not until 1965, when Mitchell visited me in New York, that I started to take his views seriously. By that time he had some impressive experimental evidence in support of his hypothesis. He had shown that both respiration and the hydrolysis of ATP are associated with the movement of protons. In Figure 5-1 the basic formulation of his hypothesis is shown. The function of the respiratory chain is to establish an electrochemical proton gradient. The function of the ATPase is to use this gradient for the generation of ATP from ADP and P_i. It operates as a reversible ATP-driven H^+ pump, or more

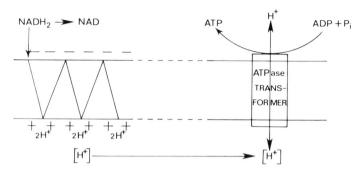

Fig. 5-1 Chemiosmotic mechanism of mitochondrial ATP generation.

appropriately as an ATP synthetase acting as a proton turbine. It generates ATP when it is provided with an electrochemical proton potential $\Delta \mu_{H^+}$ which consists of two components, the pH gradient (Δ pH) and the membrane potential ($\Delta \psi$).

It is clear that a chemiosmotic mechanism is operative in reconstituted systems. We have shown that an electrochemical proton gradient generated by bacteriorhodopsin can be used by the mitochondrial proton pump to generate ATP from P_i and ADP (Racker and Stoeckenius, 1974). Are mitochondria, chloroplasts, and bacteria, which use electron transport to generate Δ pH and $\Delta \psi$, more complex? Perhaps they are, but more likely with respect to regulation and efficiency rather than with respect to the basic mechanisms outlined above. Many controversies still need to be resolved, such as the mechanisms of action of the ATP synthetase, and of the proton pump associated with cytochrome oxidase and with other segments of the respiratory chain, and the identity of the hydrogen carriers. But the basic principle of energy transduction by an electrochemical proton gradient has been adequately established (Boyer *et al.*, 1977).

In the center stage of this process is F_1F_0, the ATP synthetase. The reconstitution of F_1F_0 into liposomes by Kagawa and his collaborators (cf. Kagawa, 1972, Racker, 1976; Kagawa *et al.*, 1982) and the demonstration that it acts as a reversible proton pump laid the experimental foundation for its role. F_1F_0 is a multi-protein complex that consists of three morphological entities:

1. A water-soluble protein called F_1, which contains the catalytic site responsible for the hydrolysis of ATP and its formation from ADP and P_i.

2. A connecting stalk between F$_1$ and the membrane which can be seen in electron micrographs but has not as yet been clearly identified with one of three polypeptides (F$_6$, OSCP, and the δ subunit of F$_1$ or CF$_1$) known to be required for the attachment of F$_1$ to F$_0$.

3. A hydrophobic part, called F$_0$.

I. F$_1$ (Mitochondrial MF$_1$, Chloroplast CF$_1$, and Bacterial BF$_1$)

A. General Comments

There are numerous and excellent reviews on the properties of F$_1$ and its subunits from mitochondria (Cross, 1981; Penefsky, 1979), chloroplasts (McCarty, 1979; Baird and Hammes, 1979; Nelson, 1981a), and bacteria (Kagawa et al., 1979; Kagawa, 1980; Futai and Kanazawa, 1980, 1983; Fillingame, 1981; Senior and Wise, 1983). I shall not dwell on many of the aspects dealt with adequately in these reviews, some of which I have discussed previously (Racker, 1976, 1981a); instead I shall concentrate on recent information derived from reconstitution experiments and studies of bacterial mutants (Dunn and Heppel, 1981; Philosoph et al., 1981; Kanazawa and Futai, 1982; Gibson, 1983; Senior and Wise, 1983). The introduction of genetic manipulations with microorganisms has given this field of bioenergetics a new dimension.

Before advancing to a detailed discussion of recent developments, a few reservations about interpretations of these experiments need to be emphasized. When we study the mode of action of an enzyme with the aid of an inhibitor, it is often difficult to decide whether the action is at the catalytic site or at an allosteric site, resulting in a conformational change of the enzyme. It was observed in the case of mitochondrial F$_1$ (Ferguson et al., 1975) and chloroplast CF$_1$ (Deters et al., 1975) that 7-chloro-4-nitrobenz-2-oxa-1,3-diazole (NBD-Cl) inhibits ATPase activity by interacting with a tyrosine residue of the β subunit. It was shown that NBD-Cl induces major conformational changes in F$_1$ (Holowka and Hammes, 1977). It does not appear to interact with the active site, since it does not interfere with the binding of 5′-adenylyl-β,γ-imidophosphate (AMP-PNP). In contrast, inhibitors such as phenylglyoxal, DCCD, and efrapeptin, interfere with the binding of AMP-PNP (Cross and Nalin, 1982). One can always raise objections to such interpretations because of the obvious complexity of the active center that is formed by several amino acid residues which offer multiple binding sites. Nevertheless, the evidence discussed above and the demonstration of another tyrosine residue (Esch and Allison, 1979), which is a more likely candidate, appears to eliminate the NBD-Cl–reactive tyrosine as a member of the active center. Another interesting complication arises from the important investigation of Penefsky and collaborators (Cross *et al.*, 1982), which I shall return to at the end of this lecture. They have shown that occupancy of a second site of F_1 by ATP leads to a 10^6-fold increase in the rate of products released from the first site occupancy and a 30-fold increase in ATP hydrolysis. Thus, small changes in assembly of subunits could dramatically alter cooperativity and rates of catalytic activity. These studies give us not only new insights into the mode of action of F_1 but also raise questions with regard to interpretations. More importantly, they open up new experimental approaches to the analysis of bacterial mutants that show defects in ATPase and ATP synthetase activity.

I have discussed the NBD-Cl site and the cooperativity studies mainly to illustrate the difficulties encountered in attempts to penetrate into the inner sanctum of a catalytic center. Similar reservations pertain to the interpretation of experiments based on reconstitution of subunits and analyses of experiments with mutated F_1. Here it is difficult to differentiate between a direct role of a subunit in a catalytic or a regulatory function on the one hand, and an indirect role in the assembly of a functional complex on the other hand. For example, it is clear now that the α, β, and γ subunits of F_1 are required for the assembly of an active ATPase enzyme; however, it appears that an intact γ subunit is not required for the hydrolysis of ATP catalyzed by CF_1 (Deters *et al.*, 1975). It was shown in these studies that CF_1, which was extensively digested by trypsin, revealed on analysis by SDS–PAGE only α and β subunits. Yet the preparation catalyzed the hydrolysis of ATP at a rate that was actually greater than that catalyzed by undigested CF_1. It was pointed out by McCarty (1979) that a fragment of the γ subunit may have remained attached to the $\alpha\beta$ complex of CF_1 and fulfilled its function. Alternatively we may be dealing with a phenomenon similar to that described for the acetylcholine receptor. Nicking of the γ subunit by trypsin may have left the $\alpha\beta\gamma$ complex assembled as a complex but the γ subunit undetectable by SDS–PAGE analysis. Although such a possibility cannot be neglected, the above mentioned increase in specific ATPase acitivity following digestion favors the physical removal of most of the γ subunit and of other minor components that are not required for ATP hydrolysis. Consistent with this view is the observation, to be discussed later, that trypsinized CF₁ is resistant to inhibition by the ε subunit.

In addition to the problem of changes in the capacity of subunit assembly, a mutational event may render a subunit less stable *in vivo*, e.g., more susceptible to proteolytic digestion, thereby obscuring the interpretation of functional changes. This possibility needs more attention than it has been given in the current literature.

B. Role of Individual Subunits

With these reservations in mind, it is possible to arrive at tentative conclusions regarding the role of the individual subunits. The reconstitution of ATPase activity with α, β, and γ subunits is now well established (Kagawa, 1978; Dunn and Heppel, 1981). Of particular interest are recent observations on species specificity. It was shown (Takeda *et al.*, 1982) that the β subunit of *E. coli* F₁ could be reconstituted with the α and γ subunits of the thermophilic bacterium PS3 and that the γ subunit from one source could be reconstituted with the α and β subunit from the other source. However, the α subunit showed some specificity in failing to reconstitute with β and γ from the alien source.

Reconstitution experiments with subunits from functionally impaired *E. coli* mutants were successfully used to define the location of the defect. For example, UncA401 has a defective α subunit. A mutation in the β subunit was discovered in UncD-11 (Kanazawa *et al.*, 1980) and was shown to alter cation specificity as well as the stability of the ATPase activity. Mutations in the γ-subunit lead to failures to form the ATPase complex in the *E. coli* membrane (Kanazawa and Futai, 1982). It was shown, however, with lambda Unc transducing bacteriophages (Mosher *et al.*, 1983) that only limited conclusions can be drawn from genetic complimentations alone, since in some instances of Unc mutants no difference could be detected between the normal and abnormal gene.

The α subunit appears to be a carrier of firmly bound adenine nucleotide, while the β subunit contains the binding site of catalytically active nucleotides (Dunn and Heppel, 1981). One role for the γ subunit is that of an organizer of assembly and, perhaps more importantly, for control of subunit interactions and of proton flux (Kagawa, 1980). The δ subunit appears to be the link between the γ and ε subunit of F_1 and F_0, the membrane sector. An antibody against the γ subunit of CF_1 is an effective inhibitor of photophosphorylation (Nelson *et al.*, 1973), perhaps by immobilizing its conformation. Revealing studies with cross-linking maleimides (McCarty, 1979) also point to a role of the γ subunit in the control of proton fluxes.

Sequential removal of the β and γ subunits of F_1 from isolated chromatophores of *Rhodospirillum rubrum* by treatment with lithium salts has yielded results that are somewhat at variance with the observation on *de novo* reconstitutions with individual subunits. Removal of β and γ subunits leaves the light-induced uptake of protons almost intact, suggesting that the γ subunit may not be essential for the gating of the proton channel, and points to a role of the α subunit in this process (Philosoph *et al.*, 1981). However, such resolution experiments are subject to criticisms that are mirror images of the objections raised to reconstitution experiments. For example, removal of a subunit from F_1 may result in a collapse and rearrangement of the residual subunits, thereby blocking an induced leakage of protons. Indeed, proton leaks in defective membranes can be repaired by addition of catalytically inactive F_1 or even by DCCD (cf. Racker, 1965, 1976) and a slow spontaneous closure of a proton leak has been observed in chloroplasts (Nelson, 1981a).

There are several experimental approaches that could shed new light on the role of the γ subunit. The question of the presence of nicked γ subunit in trypsin-digested CF_1 could be settled by appropriate cross-linking experiments prior to examination by SDS–PAGE or by analysis of the small peptides released by SDS treatment. An immunological search could help to discover γ subunit remnants.

There can be little doubt that the γ subunit plays a major role in the reconstitution of an active F_1 with α and β subunits, and probably also in the biological assembly during biogenesis (Nelson, 1981a). The answer to its role in controlling proton fluxes must come from reconstitution experiments. The pioneering experiments of Kagawa and his collaborators on TF_1 from the thermophilic bacterium PS3 (Kagawa, 1980) have shown that a mixture of isolated γ, δ, and ϵ subunits of TF_1 added to reconstituted F_0 inhibited the proton flux through the membrane. These important findings need to be extended to an analysis of the role of each of these three subunits. Kagawa's suggestion for the role of the γ subunit as a "gate" for proton flux is a provocative proposal, and we need to search further for information on how this gate is opened and closed. A possible involvement of an SH–SS interchange in the control of proton fluxes has been suggested (Alfonso *et al.*, 1981; Kantham *et al.*, 1984), but the bacterial γ subunit does not contain an SH group, whereas other subunits of the ATPase complex do. Perhaps the mechanism of F_1F_0 operation in thermophilic bacteria is less complicated than in chloroplasts and mitochondria, as is indicated by its simpler subunit composition. As suggested earlier (Racker, 1981a), in chloroplasts, mitochondria, and some photosynthetic bacteria there may be a double gate: one in F_1 and one in F_0. This could explain the observations of Philosoph *et al.* (1981) that removal of the γ subunit from chromatophores leaves the gate "almost" intact. We see throughout evolution the appearance of greater complexity and of additional functional SH groups which render the structures more delicate, but perhaps also more capable of operating with greater efficiency and subject to more rigid regulation.

The role of the two smallest subunits (δ and ϵ) of F_1 have also been elucidated by reconstitution experiments. The first resolution of the δ subunit of CF_1 was reported by Nelson and Karney (1976) and further studied later (Younis *et al.*, 1977; Andreo *et al.*, 1982). In the latter study it was established that δ-free CF_1 did bind weakly to CF_1-depleted particles but did not block the proton leak through F_0. For tight binding and blockage of the proton leak both δ and ϵ were required (Richter *et al.*, 1984). In the case of *E. coli*, both δ and ϵ participate in the binding of EF_1, but their precise contribution to the blocking of the proton leak has not been established. Since there is no evidence for OSCP in *E. coli*, the δ subunit may play a more substantial role in F_1 attachment to the membrane and in communication with the F_0 channel than in higher organisms. The observed sequence homology between OSCP and the *E. coli* δ subunit suggests such a role (Walker *et al.*, 1982; Futai and Kanazawa, 1983).

The role of the ϵ subunit is unambiguous in chloroplasts. It serves as a regulatory subunit that inhibits ATPase activity (Nelson *et al.*, 1972). Inhibition by addition of ϵ subunit to ϵ-depleted CF_1 was prevented either by an antibody against the γ subunit or by removal of the γ subunit by tryptic digestion. Addition of ϵ subunit to trypsinized $\alpha\beta$ ATPase was also ineffective. These observations point to an interaction between the ϵ and γ subunits. They also suggest reconstitution experiments of adding γ and δ subunits to the $\alpha\beta$ ATPase that could give information on the presence of a nicked γ

subunit. They could also shed light on the concept of the γ subunit as an "organizer" that interacts with multiple subunits, as indicated by cross-linking experiments (Baird and Hammes, 1979).

The metabolic role of the "ϵ subunit" in *E. coli* F_1 (EF_1) and mitochondria is more ambiguous than in chloroplasts and has, moreover, been clouded by nomenclature. In both EF_1 and MF_1 preparation an "ϵ subunit" has been described that does not seem to regulate membranous ATPase activity. Although the ϵ subunit in EF_1 participates in EF_1 binding to the membrane (and inhibits ATPase activity of soluble EF_1), it does not interfere with the hydrolysis of ATP catalyzed by membrane-bound EF_1 (Smith and Sternweis, 1977; Futai and Kanazawa, 1980). The situation is even more complex in bovine heart MF_1, which contains both a mitochondrial ATPase inhibitor (Pullman and Monroe, 1963) and an "extra ϵ subunit." Nelson (1981b) therefore argues in favor of a 6 subunit structure for F_1 with an inhibitor subunit (ϵ) and a second ϵ subunit (ϵ').

It seems appropriate at this point to discuss briefly the role of the mitochondrial and chloroplast inhibitor as elucidated by reconstitution experiments. As mentioned above, its role in chloroplasts is less ambiguous than in mitochondria and bacteria, since it is a regulatory subunit of CF_1 which can be dissociated only by harsh treatment (Nelson *et al.*, 1972; Richter *et al.*, 1984). In the case of bovine heart MF_1, the inhibitor can be dissociated by treatment with salt or gentle heat (Horstman and Racker, 1970), and the "subunit" is actually lost during the course of some purification procedures.

A regulatory role for the ATPase inhibitor has been emphasized by Ernster and his collaborators (Gomez-Puyou *et al.*, 1979) and by others (van de Staat *et al.*, 1973; Harris and Crofts, 1978; Racker, 1976). Of particular interest is the concept that the interaction of the inhibitor and F_1 is under the control of the membrane potential, which may help to explain the inhibition of ATP hydrolysis without affecting oxidative phosphorylation. Students well indoctrinated in the laws of thermodynamics find the idea of inhibition in one direction (ATP hydrolysis) without inhibition of the reverse direction (ATP generation) unpalatable. Professors with seniority and wisdom (e.g., Senior and Wise, 1983) are accustomed to obstacles placed by nature in the path of investigators and are familiar with pitfalls encountered in experiments in which kinetic properties of the systems appear to contradict the holy laws of thermodynamics. We resolved such an apparent contradiction many years ago (Racker *et al.*, 1959) when we showed that (threose 2,4-diphosphate), an inhibitor of glyceraldehyde-3-phosphate dehydrogenase, inhibits the forward but not the reverse reaction catalyzed by this enzyme. First we demonstrated that threose 2,4-diphosphate is a suicide inhibitor (probably the first one described) by serving as a substrate which is oxidized to a stable acyl enzyme intermediate. The substrate of the back reaction, 1,3-diphosphoglycerate, displaced this intermediate, thus eliminating the inhibition. The situation in the case of the mitochondrial inhibition is more complex, but it is not difficult to see how changes in Δ pH and $\Delta \psi$ could profoundly influence the affinity of the inhibitor to F$_1$ and thereby change the catalytic properties of the enzyme.

What is the physiological role of the inhibitor? Clearly, biological pumps and turbines should not be leaky if they are meant to operate efficiently. Yet MF$_1$ is clearly a very leaky enzyme in the absence of inhibitor. Nevertheless, in a system reconstituted with an active ATPase we can avoid the wasting of ATP by either adding mitochondrial ATPase inhibitor or quercetin (Lang and Racker, 1974) or by competing for ATP with an excess of hexokinase. In *E. coli*, where ATP synthesis can barely keep up with the demand for ATP for growth, an efficient system for ATP consumption is provided and no inhibitor is needed. Moreover, the proton pump in *E. coli* may be used to expel protons, e.g., during anaerobic glycolysis. An inhibitor may be undesirable under these circumstances. In mammalian cells, on the other hand, when the potential of ATP generation is far in excess of demands, both control of ATP generation and of ATP by hydrolysis are required. The inhibitor participates in this control. Why is the inhibitor readily dissociated from MF$_1$ but not from CF$_1$? Perhaps there is a biological reason for this. In mitochondria the supply of oxidizable substrate is continuous, and there is a need for the ATP-driven formation of NADPH used for many biosynthetic processes, and flexibility of inhibitor reactivity is required. In chloroplasts the light-dependent ATP formation is discontinuous and ATP needs to be preserved. In the dark, because of the presence of inhibitor, there is little ATPase activity detectable. In the light, illumination induces ATPase activity under some experimental conditions (Mills *et al.*, 1980).

Why do the ATPase-driven proton pumps of mitochondria and chloroplasts have a dissociable inhibitor which appears to be lacking in other ATP-driven pumps? Perhaps the answer to this question can be found in the curious fact that F_1, the transformer of oxidative energy, is located outside the membrane in a hostile hydrophilic environment. As will be discussed in Lecture 7, other pumps have hydrophobic devices within the membrane that permit efficient function.

I shall not deal here with details of the sizes, chemical properties, and chemical alteration of F_1 subunits, but will refer to the excellent review by Futai and Kanazawa (1980). Nor will I discuss the question of subunit composition of F_1, except to say that the evidence is clearly swinging the pendulum towards a $\alpha_3\beta_3\gamma\delta\epsilon$ composition (Yoshida *et al.*, 1979; Dunn and Futai, 1980; Senior and Wise, 1983).

Based on immunological, cross-linking, nucleotide binding, and reconstitution studies discussed earlier, the arrangement of F_1 subunits as $\alpha\beta\alpha\beta\alpha\beta$ is pictured. An alternative model with an $\alpha\alpha\alpha\beta\beta\beta$ assembly, based on X-ray diffraction studies, was presented by Pedersen (1982).

The proton channel is visualized to reach the catalytic center formed between α and β via a helical δ under the control of conformational changes in the γ subunit.

This description of the subunits of F_1 and their function requires some expansion and recapitulation (Table 5-1).

The Subunits of F_1

1. All F_1 species and their subunits are created equal, but some are more equal than others. Bacterial F_1 is the simplest; mitochondrial F_1 the most complex.

2. The α and β subunits play the central role in the binding and catalytic conversion of P_i and ADP to ATP in all F_1 species. Confor-

TABLE 5-1 Functions of Subunits of F_1 and CF_1

$\alpha + \beta$ =	ATPase
α =	Part of active site; has ATP high affinity site and control function
β =	Part of active site for ATP hydrolysis and synthesis
γ =	Required for assembly of $\alpha + \beta$ and interaction with small subunits; control of proton flux
δ =	Attachment to membrane (part of stalk)
ϵ	Contribute to attachment
ϵ'	Control of ATP hydrolysis

mational interaction between α and β appears to be required for function (Senior, 1985).

3. Confusing differences described in the literature, e.g., in the case of the ϵ subunit, are partly semantic, partly due to differences in physiology. We can explain why bacteria do not need an ATPase inhibitor, why mitochondria need a loosely bound and chloroplasts a tightly bound inhibitory subunit.

4. The δ subunit probably plays the role of the F_1 proton channel with the γ subunit as the gate. The reality is probably more complex.

5. In reconstitution experiments, the subunits also have a structural role governing the assembly of the F_1 complex.

II. The Stalk, OSCP, and F_6

Between F_1 and the membrane there is a stalk. What are the biochemical equivalents for the morphological presence of the stalk? Is it real or is it an artifact of drying and negative staining required for visualization by electron microscopy?

A. Evidence for the Reality of the Stalk

In view of still existing doubts expressed in reviews about the location of F_1 outside the membrane and the existence of the stalk, I shall summarize what I believe to be persuasive evidence.

1. In addition to the demonstration of F_1 by negative staining, its location outside the membrane can be clearly seen after fixation of mitochondria or chloroplasts with glutaraldehyde by positive staining with methanolic osmium tetroxide (Telford and Racker, 1973). Previous failures to see the protein in positive stains, which have contributed to the myth of its intramembranous location, were caused by insufficient penetration of aqueous osmium tetroxide. Similar pictures of positively stained chloroplasts have been published by Oleszko and Moudrianakis (1974).

2. Electron micrographs of freeze-etch preparations of chloroplasts show the characteristic spheres of CF_1. The identity of the particles was demonstrated by resolution and reconstitution of CF_1 to depleted chloroplasts (Garber and Steponkus, 1974).

3. Shadowed freeze-etch preparations of submitochondrial particles allow for quantitative estimates of the distance of F_1 from the

membrane (Telford *et al.*, 1984). By standardizing the measurements of shadows with gold particles, it was shown that over 80% of the CF_1 molecules protrude by 134 Å from the surface. This value is very close to the sum of the F_1 diameter of 96 Å and of the estimated 35–40 Å length of the stalk.

4. The accessibility of CF_1 antibodies against its individual subunits makes it highly improbable that the protein is buried in the membrane (Nelson *et al.*, 1973).

5. Experiments were performed with radioactive diazobenzene sulfonate (Schneider *et al.*, 1972). This impermeant but rather unspecific reagent was shown to react quantitatively to the same extent with F_1 solution and with F_1 attached to phosphorylating submitochondrial particles. In contrast, cytochrome oxidase treated with the same reagent in submitochondrial particles contained less than 20% of the label found in solubilized cytochrome oxidase exposed to the reagent.

The extramitochondrial location of F_1 is of considerable significance, in view of its generally recognized function as the energy transformer of proton flux-dependent phosphorylation. It assigns to the stalk the responsibility of proton flux transmission. OSCP was earlier considered to be the logical candidate for the stalk (Senior, 1973) because it was a component linking F_1 to the membrane. However, the ATPase complex of chloroplasts does not appear to have an OSCP equivalent, yet shows a stalk in electron micrographs. The most likely candidate for the stalk is the extruded δ subunit of F_1, which in fact is visible in a rudimentary form in electron micrographs of F_1 from insects (Younis *et al.*, 1978). Possible contributions to the stalk by the ε subunit of CF_1 and BF_1, by the ε' subunit of F_1, by the subunit b of BF_0, as well as by OSCP and F_6, need to be considered. Attempts to reconstitute the stalk from individual proteins have thus far not been successful but need to be continued. When this has been accomplished, a *de novo* assembly of F_1F_0 from individual polypeptide chains could be more readily achieved.

B. The Oligomycin-Sensitivity Conferral Protein (OSCP)

The protein was first discovered in crude alkaline extracts of mitochondria as a component required for the conferral of oligomycin-sensitivity to F_1 (Kagawa and Racker, 1966) and was purified by MacLennan and Tzagaloff (1968). OSCP interacts with F_1, probably at a ratio of $3:1$, and it binds more tightly to it than to F_0 (Hundal and Ernster, 1979). It should be emphasized that OSCP (in spite of its name) does not interact with oligomycin but is required for the binding of F_1 to the hydrophobic sector (Knowles et al., 1971). Reconstitution experiments suggest that it may have a more specific function in addition to its structural role as a connecting link. A membranous ATPase complex deficient in OSCP as well as in phospholipids catalyzed ATPase activity that was not sensitive to oligomycin. Most noteworthy is the fact that addition of OSCP completely blocked ATP hydrolysis. Phospholipid addition restored the activity which was now sensitive to oligomycin or DCCD (Bulos and Racker, 1968). We therefore suggested that OSCP may participate in the opening and closing of the proton channel together with other components of F_0 (Alfonso and Racker, 1979) in a mechanism of cyclical proton fluxes proposed for ATP synthesis (Racker, 1977a,b).

C. The Heat-Stable Coupling Factor F_6

F_6 together with OSCP is required for the "proper" binding of F_1 to the hydrophobic sector of the ATPase complex, since neither alone is sufficient for the reconstitution of a functional F_1F_0 complex (Knowles et al., 1971). F_6 was isolated in homogenous form (Kanner et al., 1976) and was sequenced (Fang et al., 1984). Experiments on the sensitivity of F_6 to trypsin suggest that this protein also has a functional role in addition to a structural one. It appears that its structural role (as an attachment factor for F_1) is more resistant to trypsin than its function in the reconstitution of the proton pump.

D. Coupling Factor 2 (Factor B)

This protein stimulates the $^{32}P_i$–ATP exchange in silicotungstate-treated submitochondrial particles in the presence of excess F_6 (Racker et al., 1969). F_2 is identical with factor B (Sanadi et al., 1968; Racker et al., 1970) and is highly sensitive to sulfhydryl-modifying reagents (Racker and Horstman, 1967; Sanadi et al., 1968). Many of its properties have been described (Hughes et al., 1979;

Kantham *et al.*, 1984), but its precise mode of action still needs to be elucidated. It is significant that a preparation of F_1F_0 that contained large amounts of factor B had a very high $^{32}P_i$–ATP exchange activity (Hughes *et al.*, 1982).

III. The Hydrophobic Sector

A. *Composition of the Hydrophobic Sector*

It appears that the ATPase complex has the largest number of required subunits in mammalian mitochondria and the smallest in thermophilic bacteria.

1. Mitochondria

In bovine heart mitochondria, in addition to the three factors (OSCP, F_6, and F_2) that can be rendered water-soluble in the absence of a detergent, there are several additional components that constitute the hydrophobic sector of F_0 and are firmly embedded in the phospholipid bilayer. There is evidence from reconstitution studies with bovine heart F_0 for at least two hydrophobic components that are required for function. One is the DCCD-reactive proteolipid; the second is a 28,000-dalton protein (Alfonzo and Racker, 1979).

In yeast mitochondria a similar complexity is found (Tzagoloff, 1982). Moreover, a second proteolipid with a molecular weight of 5870 has been recently implicated as a component of the ATPase complex (Velours *et al.*, 1984).

2. Chloroplasts

The composition of CF_0 appears simpler. There is no evidence for the participation of OSCP or F_6. There are three hydrophobic subunits (Nelson 1981a,b) that have been named I (15 kDa), II (12.5 kDa), and III (8 kDa). The latter is the DCCD-reactive proteolipid which I shall discuss in greater detail. Subunit I has been proposed to serve as a binding site for CF_1 and subunit II as a channel organizer for the proteolipid subunits (Nelson, 1981a).

3. Bacteria

A wealth of new information about F_0 in *E. coli* has emerged from the study of mutants (Gibson, 1983) and genetic analyses. There are three genetically identified subunits of F_0 in *E. coli*. The sequences of the genes have been determined in several laboratories (Gay and Walker, 1981; Futai and Kasanawa, 1983; Nielson *et al.*, 1981). Subunit a has a molecular weight of 30,000 and is a very hydrophobic protein; subunit b has a molecular weight of 17,200 and functionally resembles in some respects OSCP; and subunit c (molecular weight of 8300) is the DCCD-binding proteolipid. The stoichiometry of these subunits is probably close to $1:2:10$ (Foster and Fillingame, 1982). Topology and reconstitution analysis of this complex are being conducted (Perlin *et al.*, 1983; Hoppe *et al.*, 1983). A highly purified, reconstitutively active preparation of F_0 with a similar composition of three subunits was described by Schneider and Altendorf (1982; 1984). Some pitfalls, in addition to those mentioned earlier, need to be emphasized, since they are apparently not fully appreciated. The use of proteases in the analysis of the topology of F_1F_0 was first applied to bovine submitochondrial particles (Racker, 1963) and more recently to F_0 of *E. coli* (Hoppe *et al.*, 1983). They are useful approaches, but interpretations are not without ambiguity. Some proteins attached to membranes show not only remarkable resistance to proteases but also protect subunits that are sensitive. The allotopic changes in susceptibility to trypsin have been discussed earlier. The functional changes induced by proteolytic digestion need to be reemphasized here, since they pertain to reconstitution experiments with defective subunits from mutants. We had observed many years ago (Horstman and Racker, 1970) that exposure of F_1 to proteases did not interfere with its ability to bind to the membrane and even to retain some rutamycin sensitivity, but completely abolished its coupling factor activity. A similar differential effect of trypsin on F_1 has been also observed in *E. coli* F_1F_0 (Dunn *et al.*, 1980). These curious observations have never been satisfactorily explained. What they have taught us, however, is that a binding assay and even conferral of sensitivity to energy-transfer inhibitors are not ultimate assays for reconstitution. On the other hand, I want to stress the great usefulness of oligomycin and other energy-transfer inhibitors discovered mainly in the laboratory of Lardy *et al.* (1958) (Lardy, 1980). These inhibitors were instrumental in the discovery of F_0 (Racker, 1963) and indispensible during purification of the hydrophobic subunits (Alfonzo and Racker, 1979).

Another complexity of F_1-binding assays was observed by Bulos and Racker (1968). F_1 binding to membranes depleted of coupling factors took place without OSCP, provided sufficient cations were present (60 mM K^+ or 2 mM Mg^{2+}), but the ATPase activity of the bound enzyme was insensitive to energy-transfer inhibitors. Similar observations with $E.$ $coli$ F_1 treated with trypsin (Hoppe et $al.$, 1983; Perlin et $al.$, 1983) require further analyses. Apparent discrepancies of observations between the two laboratories might be resolved by studies of the different types of binding previously described for bovine F_1.

Lesson 19:   On the various types of attachments of F_1

These are various kinds of attachments between F_1 and membrane components:

A readily dissociable friendship, such as the binding of F_1 to F_0 without the glue of OSCP and phospholipids; or the attachment of δ-free CF_1 to CF_1-depleted chloroplast membranes.

An unproductive association with animosity, such as F_1 and OSCP added to phospholipid-depleted F_0, resulting in inhibition of ATP hydrolysis activity.

An unproductive love affair between F_1 which has a personality defect (such as trypsin-treated F_1) and the membrane, resulting in no coupling activity.

An engagement lacking housing facilities, such as seen in an oligomycin-sensitive ATPase containing insufficient phospholipids to provide a vesicular compartment.

Marriage, a productive chaining of F_1 to a channel with energetic consequences.

The simplest reconstitutively active F_0 has been described in thermophilic PS3 bacteria. In addition to the DCCD-reactive proteolipid, only one hydrophobic component of 13,500 M_r was shown to be required for the reconstitution of vesicles mediating all F_0 activities (Sone et $al.$, 1978). This system should be most suitable for the elucidation of the channel mechanism.

B. The DCCD-Reactive Proteolipid

This is the most extensively studied component of the hydrophobic sector. Its chemical properties have been reviewed (Kagawa,

1980; Futai and Kanazawa, 1980; Sebald, 1977; Sebald and Hoppe, 1981), and there is general agreement that it functions as a proton channel. The most persuasive evidence for its function was obtained by its isolation from chloroplasts in a form suitable for reconstitution into lipid vesicles (Nelson *et al.,* 1977; Sigrist-Nelson and Azzi, 1980). The success of these experiments was predicated on the remarkable stability of the chloroplast proteolipid in butanol. Similar successful experiments with butanol have been recently described with the proteolipid from yeast (Nelson, 1981b). Unfortunately the same butanol extraction procedure has failed with bovine heart mitochondria (N. Nelson, unpublished observation) and in thermophilic bacteria (Kagawa, 1980).

Although the importance of the successful experiments can hardly be overemphasized, it should be realized that they are suffering from several deficiencies. The rates of proton conduction are much lower than expected, and relatively high concentrations of DCCD are required to block the proton flux (Nelson, 1981a,b). Moreover, the butanol-treated proteolipid has thus far not been reconstituted with the other components of F_1F_0 to yield an active proton pump. Imaginative interpretations of these findings, involving another polypeptide helping in the assembly of the channel into a hexameric (or larger) polymer, have been presented (Nelson, 1981b), but the future will tell whether they will prove to be either correct or at least the springboard for future experiments. I shall come back to the role of other hydrophobic polypeptides in F_0.

The proteolipid is resistant to proteolysis but remarkably sensitive to solvents such as chloroform-methanol. Reconstitution of the channel with a proteolipid extracted from yeast mitochondria with chloroform-methanol has been claimed (Criddle *et al.,* 1977) but shown to be probably an artifact, since exclusively voltage dependent K^+ conduction can be observed (Nelson, 1981b). In view of the sensitivity of proteolipids to chloroform-methanol (see also Knowles *et al.,* 1980), one wonders whether even butanol treatment yields a native protein, particularly since attempts to reconstitute such preparations with F_1 have thus far failed.

We have taken advantage of the resistance of the proteolipid to proteolysis to separate it from other hydrophobic components of bovine F_0 (Alfonso and Racker, 1979). Although treatment with trypsin did not yield a homogeneous preparation of the proteolipid, it led to the discovery of a second hydrophobic component of 28,000 dalton which is highly sensitive to trypsin and resistant to solvents.

C. The Second Hydrophobic Component of F_0

Information on the second hydrophobic component of F_0 is very incomplete. In bovine heart F_0, the properties of a partially purified 28,000-dalton component have been described (Alfonso and Racker, 1979; Alfonso et al., 1981). It is very sensitive to trypsin and is inhibited by interaction with p-hydroxymercuribenzoate. We assume that it is located at the M-side of the inner mitochondrial membrane at the orifice of the proton channel because of its susceptibility to inactivation by trypsin in F_1-depleted submitochondrial particles. We have proposed (Alfonso et al., 1981) that it serves as the connecting link between F_6 and OSCP and the DCCD-reactive proteolipid. It could serve as the "channel assembly protein" proposed by Nelson (1981b). We have suggested that the 28,000-dalton protein acts as a valve (F_0 ring) participating in the closing and opening (or assembly and disassembly) of the channel, a function essential for a hypothesis of a mechanism of energy transduction involving Mg^{2+} binding as the central theme (Racker, 1977a,b). The weaknesses of these propositions, in addition to the fact that the hypothesis is based on experiments conducted with the Ca^{2+} ATPase of sarcoplasmic reticulum, are failures to detect a 28,000-dalton component in F_1F_0 preparations from thermophilic bacteria (Kagawa, 1978) or from chloroplasts (Pick and Racker, 1979a). However, both of the two latter F_0 preparations contain an unidentified hydrophobic component of about 14,000 daltons. Its function is not clear, and experiments on chloroplast F_0 with p-hydroxymercuribenzoate have not revealed sensitivity to this reagent (A. Kandrashin, A. Kandrach, and E. Racker, unpublished observations). Subunit a of E. coli (molecular weight of 30,000) is very hydrophobic and may be the equivalent to the 28,000-dalton protein. However, it was found to be quite resistant to proteolysis (Perlin et al., 1983). Attempts to obtain homogeneous 28,000-dalton protein from bovine F_0 have been hampered by its hydrophobicity and tendency to aggregate. I mentioned earlier that a particulate preparation of F_1F_0 deprived of both OSCP and phospholipids catalyzes ATP hydrolysis resistant to DCCD or rutamycin. OSCP completely inhibited the ATPase activity, which was restored by further addition of certain phospholipids and thereby rendered rutamycin-sensitive (Bulos and Racker, 1968). We have obtained data showing that the trypsin-sensitive 28,000-dalton protein is a participant in the inhibition observed on addition of OSCP (Alfonso and Racker, 1979). There have been reports from other laboratories both on the presence or absence of a protein with an M_r of 30,000–32,000 in preparations that catalyze oligomycin-sensitive $^{32}P_i$-ATP exchange (Bäumert et al., 1981; Berden and Henneke, 1981). Only the latter reported on reconstitution experiments and observed a partial reconstitution of oligomycin-sensitivity by addition of serum albumin. No data on inhibition of ATPase activity were recorded.

D. Formation of the Channel

The complex interactions between the components of F_1F_0 can be explained by the following series of proposed events. In the absence of phospholipid, the channel formed by the proteolipid is closed by the 28,000-dalton protein, which serves as a valve. Added F_1 is bound to the complex by means of F_6 but is not properly linked to the channel. Oligomycin-insensitive ATPase activity is expressed as if F_1 were in solution. On addition of OSCP, the connection with the channel is established, but the gate is closed by the 28,000-dalton protein and the protons generated by ATP hydrolysis cannot escape, resulting in ATPase inhibition. On addition of specific phospholipids there is interaction with the hydrophobic 28,000-dalton protein, the gate is opened, and rutamycin-sensitive ATPase activity is induced. The role of the phospholipids in this phenomenon needs to be explored further. The restoration of DCCD-sensitive ATPase showed interesting phospholipid specificity (Bulos and Racker, 1968). For example, phosphatidylinositol 4,5-bisphosphate activated about 70% of the ATPase of the lipid-depleted complex but did not confer rutamycin sensitivity. Phosphatidylinositol activated about 50% of the masked ATPase, which was inhibited 60% by rutamycin. On addition of soybean phosphatidylcholine, only 24% of the ATPase activity was expressed but rutamycin inhibited 63%. Thus, two distinguishable functions of phospholipids are suggested by these studies: the opening of the channel, by interaction with a hydrophobic component of F_0, and the facilitation of interaction with energy transfer inhibitors. In fact, one can formulate a mode of action of energy transfer inhibitors that requires the interaction between the 28,000-dalton protein and a specific lipid. Thus, the covalent interaction between, e.g., DCCD and the proteolipid monomer may not directly close a channel, but induces a conformational change in the proteolipid that maintains the channel in the closed (or disassembled) configuration. To explain the role of a second hydrophobic protein (e.g., the 28,000-dalton protein in bovine heart F_0) we propose the existence of two conformations of the native channel: The c (closing) conformation, which can be closed by the 28,000-dalton protein in the presence of the appropriate lipid, and the o (opening) conformation. The energy transfer inhibitor stabilizes the c conformation. According to this formulation the inhibitor does not actually close the channel but prevents its opening (or assembly). Whether the specific lipid plays a role in a cyclic alteration of conformation of the proteolipid or regulates the conformation of the 28,000-dalton protein that interacts with the proteolipid needs to be elucidated. We have proposed that the binding and release of Mg^{2+} is associated with the cyclic opening and closing of the channel (Racker, 1977a,b).

There is little information on the b subunit of *E. coli* F_0. In view of its size, the large number of hydrophilic residues, and proposed link to F_1 (Walker *et al.*, 1982; Perlin *et al.*, 1983), it probably functions like OSCP or OSCP + F_6. The recently reported successful separation of *E. coli* subunit b from the a–c complex and its successful reconstitution (Schneider and Altendorf, 1984) represents an important step toward the elucidation of its function.

> *Lesson 20: Generalizations aren't worth a damn, including this one (Oliver Wendell Holmes)*

It was said that "what is good for General Motors is good for the country." This is a narrow viewpoint looking at economic values only, and even then it is likely to be fallacious. The biological equivalent would be an equally fallacious statement that "What is good for *E. coli* is good for '*Homo sapiens*'." The eight subunits of F_1F_0 in *E. coli* serve the purposes of *E. coli*, which are different from those of mitochondrial F_1F_0. *E. coli* F_1F_0 supplies not only ATP but also can be used to expulse protons during anaerobic glycolysis. An inhibitory subunit may not be desirable under those circumstances. Moreover, in contrast to mammalian and plant cells that are capable of generating a great deal more ATP than they are likely to need except in emergencies (induced, e.g., by excessive jogging), *E. coli* uses ATP as it is made for reproductive purposes and may do without a regulatory inhibitory subunit. Mitochondria may need to go into reverse for the generation of reduced NADP, and a readily dissociable inhibitor may be a useful feature. I hope that the brilliant work on *E. coli* mutants and genes will not blind our vision to the greater complexity of mammalian structures.

IV. Mechanism of Action of F_1

Before closing this lecture I would like to draw your attention to recent experiments that shed light on the mode of action of F_1. They illustrate once more the need for resolutions and for studies of the isolated components of a multicomponent complex. Several years ago Boyer (1979) formulated the "binding-change mechanism" of ATP formation. He prefers this name to previous designations such as "energy-linked conformational mechanisms," "alternating-site mechanisms," and even "Boyer mechanisms," and we should accept his wishes. His formulations explain a number of observations on oxygen exchanges and on formation and labeling of nucleotides. The proposition is that energy-induced changes in F_1 conformation induce changes in chemical reactivity involved in the binding of P_i and ADP and release of ATP. A 30-fold increase in the rate of ATP hydrolysis when a second site of F_1 binds ATP (Cross *et al.,* 1982) is a dramatic example compatible with the "binding-change mechanism." Similar experiments were reported by Kozlov (1981). Two other significant observations have been described with pure CF_1 (Feldman and Sigman, 1982) and F_1 (Sakamoto and Tonomura, 1983). In the case of CF_1, net ATP formation from P_i and enzyme-bound ADP was reported. In the case of F_1, bound ATP was formed from medium ADP and P_i in the presence of 30% DMSO. In both cases the optimal pH was below neutrality. Somewhat less than 1 mol of ATP per mol of enzyme was formed.

I shall discuss later the "Mg^{2+}-binding mechanism" that I have proposed for ATP formation by Ca^{2+}-ATPase (Racker, 1977b). It does not seem difficult to see the analogy with F_1 experiments. In the case of Ca^{2+}-ATPase, an aspartyl phosphate is formed at pH 6.0 on addition of Mg^{2+} and P_i; in the case of CF_1 and F_1, the P_i acceptor is the β phosphate group of bound ADP. Enzyme-bound ATP is the phosphorylated intermediate of oxidative phosphorylation. This controversial subject has finally landed on the solid ground of chemistry. But we need more experiments. Since an ion gradient is not required for the formation of ATP, we need to explore how the release of ATP from the enzyme is achieved by the $\Delta \mu_{H^+}$. Attempts in this direction are being made (Kozlov, 1981).

Lecture 6

The E_1E_2 Pumps of Plasma Membranes

> The against comes
> before the for.
>
> **Pablo Picasso**

> Having precise ideas often leads to a man
> doing nothing.
>
> **Paul Valery**

E_1E_2 pumps are defined as ATP-driven transporters of ions that involve a phosphorylated intermediate, $E_1{\sim}P$, that is formed from ATP, converted to $E_2{-}P$, and then hydrolyzed to E and P_i. E_1 and E_2 are conformational variants of a protein of 100,000–150,000 daltons. In the reverse direction, $E_2{-}P$ is formed from P_i in the presence of Mg^{2+} and can be converted to $E_1{\sim}P$ under appropriate conditions as shown in a very simplified scheme (Fig. 6-1).

I. The Na^+,K^+ Pump and Na^+,K^+-ATPase

Books and many reviews have been written on the structure and kinetic properties of the Na^+,K^+-ATPase and its function as an ion pump (Skou and Norby, 1979; Cantley, 1981; Carafoli and Scarpa, 1982). Its primary functions are the maintenance of intracellular K^+ and the excretion of Na^+. The ion gradients thus created participate in turn in the Na^+-linked movement of sugars and amino acids, in the expulsion of Ca^{2+} and H^+ from the cell, in the secretion of salt

87

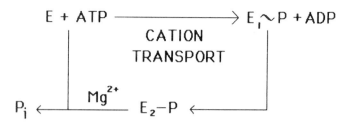

Fig. 6-1 Cation transport via E_1E_2 pumps.

from salt glands, kidney, and skin cells, and in the regulation of cell volume. It plays a particularly important role in excitable tissues, where it restores the ionic status quo after the influx of Na^+ and other ions via receptor-controlled channels. It is not surprising, therefore, that kidney, electric organs, salt glands, and brain are rich and popular sources for the isolation of this enzyme.

A. The Protein Composition of the Na^+,K^+-ATPase

There are two major subunits associated with the Na^+,K^+-ATPase which have resisted resolution without loss of ATPase activity. The large α subunit has an M_r of about 100,000, the smaller β subunit is a glycoprotein with an M_r for the protein moiety of about 40,000. There are differences in the mobilities in SDS–PAGE of both α and β subunits of Na^+,K^+-ATPase depending on the enzyme source, and even within one organ, multiple types of Na^+,K^+-ATPase can be distinguished by the size of the α subunit and the sensitivity of the enzyme to ouabain (Sweadner, 1979). This may seriously complicate conclusions with respect to subunit composition (Peterson *et al.,* 1982). The stoichiometry of subunits has been controversial, with the current pendulum swinging firmly toward an $\alpha:\beta$ ratio of $1:1$. It is of interest to note (Moczydlowski and Fortes, 1981) that some earlier and erroneous conclusions leading to controversies were based on reliance on the protein determination described by Lowry *et al.* (1951).

Lesson 21:   Quantitative determinations of proteins

It always comes as a shock to students when I tell them that the method of protein determination developed by my greatly admired friend Oliver Lowry cannot be used as a gold standard. Not for gold nor for proteins (as he clearly stated in 1951). Yet its great value (indicated by its top stock market place in the Citation Index) is fully deserved because it is a convenient and rapid method, provided all interfering buffers, detergents, and thiols are removed prior to assay. Curiously, protein estimates for several ATPases by this method are too high. For F_1 we need to divide by 1.2; for Na$^+$,K$^+$-ATPase by 1.4. Since it is impractical to determine the amino acid composition or the dry weight of samples during purification of a protein, we shall continue to use the "Lowry," the "Bradford," or even the "Warburg" 280/260 ultraviolet absorption methods of protein determination with crude preparation. Errors in both directions (low and high values) usually cancel out, and with a pure preparation we can celebrate the achievement of homogeneity by performing an amino acid analysis and standardizing the Lowry procedure for ourselves and other investigators.

There is no general agreement on the so-called γ subunit, the proteolipid which is seen in reconstitutively active preparations of Na$^+$,K$^+$-ATPase (Reeves et al., 1980). As in the case of the Ca^{2+} pump of sarcoplasmic reticulum (Knowles et al., 1980), pump activity has been observed with reconstituted ATPase preparations that have much less than one equivalent of proteolipid (Goldin, 1979). However, the Na$^+$ flux rates described are too low to allow a decivise conclusion, as will be discussed later. It should also be mentioned that proteolipids may be extracted by methanol during removal of Coomassie blue following SDS–PAGE, unless proper fixation of this protein prior to washing is achieved.

The α subunit is the energy transformer of the pump and interacts with both ATP and ouabain. Little is known about the function of the β subunit. The observations that antibodies and other agents that interact with the β subunit inhibit ATPase activity (Rhee and Hokin, 1975; Zaheer et al., 1981) does not imply a direct involvement at the active center, since either steric hindrance or secondary conformational changes may be responsible. Nevertheless, the inhibitions by anti-β subunit or by Cesalin, which interacts with the β subunit, could be useful as probes for studies of interactions between α and β subunits. The claim that phosphorylation of the β subunit influences the efficiency of pumping (Spector et al., 1980) has been withdrawn (Racker, 1981b).

The specificity of the enzyme for Na$^+$ is striking, whereas K$^+$ is replaceable by Rb$^+$ (^{86}Rb is a convenient ion for flux studies) as well as by Li^+, NH_4^+, and some other monovalent cations. Yet K^+ competes at the Na^+ site, Na^+ competes at the K^+ site, and the latter interacts with protons, a phenomenon of considerable importance for the kinetic properties of the pump *in vivo* (Skou, 1982) as well as for studies of ATPase activity *in vitro*. When ATPase activities are measured in membrane preparations that are not compartmentalized, the assay conditions are unphysiological. Both Na^+ and K^+ sites of the enzyme are exposed to unnaturally high concentrations of the opposite cation (e.g., Na^+ ion at the K^+ site) that may distort not only the rates of hydrolysis but also susceptibility to inhibitors. In unpublished experiments (E. Racker and C. Riegler, 1983), we observed that permeabilized cells that are exposed to the same ionic composition inside and outside do not catalyze a rate of ouabain-sensitive hydrolysis of ATP expected from studies of intact cells. In view of these observations and the fact that both Na^+ and K^+ interact at both sides of the membrane, it may be illuminating to study the properties of the ATP-reactive side (A-side) and the ouabain-reactive side (O-side) in reconstituted vesicles of inside out and right side out configuration under conditions of controlled ion compositions both inside and outside of the vesicles. Since neither ouabain nor ATP can diffuse across membranes, such study will require not only the use of pure enzyme preparations that are completely reconstituted, but also an intravesicular ATP-regenerating system and orientation-directed reconstitutions. I shall come back to the problems of oriented reconstitution in Lecture 12.

Most important for our understanding of the mode of action of the pump as well as for reconstitution studies is the observation that the $\alpha\beta$ monomer of the Na^+,K^+-ATPase is active (Moczydlowsky and Fortes, 1981) similar to the monomer of the Ca^{2+}-ATPase of sarcoplasmic reticulum (Dean and Tanford, 1978). However, the $\alpha_2\beta_2$ dimer appears to be present in native membranes (Kyte, 1975) and could represent the physiologically active species. Even so, the concept of the half-site reactivity mechanism based on a functional $\alpha_2\beta_2$ dimer has been made very unlikely by the penetrating studies of Moczydlowski and Fortes (1981).

In view of the similarity with the Ca^{2+} pump, it seems probable that the ion channel is formed by the α subunit of the Na^+,K^+-ATPase and undergoes conformational changes of the type proposed by Jardetzky (1966). The more specific model proposed by

Moczydlowski and Fortes (1981) for its mode of action is particularly attractive.

B. The Lipids in Na⁺,K⁺-ATPase

There is considerable information on the lipid composition of the Na^+,K^+-ATPase (DePont *et al.*, 1978). About one third is phosphatidylcholine, another third is phosphatidylethanolamine, and the last third is mainly shared by sphingomyelin, phosphatidylserine, and phosphatidylinositol. There is further information on the influence of the lipid composition on catalysis based on reconstitution experiments as well as on phospholipid modification *in situ*. Reconstitution experiments with pure phospholipids have established that the pump can function with phosphatidylethanolamine as the major lipid component of the vesicles (Racker and Fisher, 1975). Two reservations should be expressed with regard to these experiments. The rates of Na^+ fluxes of vesicles prepared by the direct sonication procedure used in this study were far below those obtained by other procedures (Hokin and Dixon, 1979; Racker *et al.*, 1979; Karlish and Pick, 1981; Cornelius and Skou, 1984). Secondly, and perhaps more importantly, in all reconstitution experiments with enzymes that have not been delipidated, phospholipids that are firmly bound to the enzyme may play a crucial part in transport activity, even though they represent a minor fraction of the total lipid. This consideration is particularly applicable when sonication in the absence of detergents is used for reconstitutions. The presence of detergents usually facilitates an exchange between the added lipid and the protein-bound phospholipids.

The Na^+,K^+-ATPase was activated with phosphatidylcholines containing fatty acids of various chain length (Johannsson *et al.*, 1981). Between 16 and 21 carbons were found to be optimal. An interesting approach to the lipid specificity of Na^+,K^+-ATPase was devised by DePont *et al.* (1978), who eliminated endogenous acidic phospholipids by exposing the enzyme to phosphatidylserine decarboxylase and to a phospholipase C, which degraded phosphatidylinositol. Since little effect was noted on catalysis, these experiments show that, at least for ATPase activity, endogenous acidic phospholipids are not required. Numerous claims have been made on the role of specific acidic lipids on the activity of Na^+,K^+-

ATPase. It is clear from reconstitution experiments that will be discussed later that acidic phospholipids are not essential for the basic operation of the E_1E_2 pumps. Yet it seems likely that secondary effects of such acidic phospholipids or even of neutral lipids, as observed in the case of the acetylcholine receptor (Killian *et al.*, 1980), may have escaped detection because of an inadequate state of purity of both enzymes and lipids used in the past and because of the rather few observations with reconstituted proteoliposomes that catalyze high ion flux rates.

C. Mechanism of Action

I shall not discuss in this lecture in any detail the molecular mechanism of action of the Na^+,K^+-ATPase as a pump. Several ingenious models have been presented by Mitchell, Boyer *et al.*, Scarborough, and others in a symposium on "Transport ATPases" (Carafoli and Scarpa, 1982). It still remains to be established whether a mobile aspartyl phosphate intermediate functions as a "miniature statistical–mechanical engine," transporting cations from one side of the membrane (Boyer, 1979; Racker, 1977a,b). The secrets of such con-the protein undergoes conformational changes that govern the binding and release of nucleotides and cations at the two sides of the membrane (Boyer, 1979; Racker, 1977a,b). The secrets of such conformational changes may indeed one day be fully "visible to the onlookers" in spite of Peter Mitchell's pessimism and objections to "conjuring bioenergetic fingers." Among the many different E_1E_2-ATPases that form an aspartylphosphate intermediate, there may be one that, because of its stability and requirement for simple lipids that do not exhibit interfering fluorescence and light scattering, lends itself to sophisticated physical–chemical analyses during transport activity, revealing comformational changes. Thus far, the simplest one is the Ca^{2+}-ATPase of sarcoplasmic reticulum, and the little I have to say about the mechanism of action of E_1E_2-ATPases will be in the section dealing with this enzyme.

An interesting model involving different conformational modifications of the Na^+,K^+-ATPase was derived from experiments exposing the enzyme to tryptic digestion in the presence of either Na^+ or K^+ (Jørgensen *et al.*, 1982). Data obtained after reconstitution of such preparations suggest that transition from $E_1{\sim}P$ to $E_2{-}P$ is required not only for the catalysis of Na^+/K^+ counterflow but also for Na–Na and Rb–Rb exchange reactions. Further reconstitution experiments with such partially crippled enzymes should yield valuable new information.

D. Reconstitution of Na^+K^+ Pump Activity

1. Freeze–Thaw Sonication

Early reconstitutions of the Na^+K^+ pump by cholate dialysis (Goldin and Tong, 1974; Hilden *et al.*, 1974) and sonication (Racker and Fisher, 1975) yielded vesicles with low transport activities. The more rapid procedures of freeze–thaw sonication (Hokin and Dixon, 1979) and freeze–thaw sonication in the presence of cholate (Karlish and Pick, 1981) yielded considerably higher rates of Na^+ fluxes (1.0–1.5 μmol per min per mg protein). In these and most other reports, calculations were based on "total protein added" rather than on mg protein of inside-out-oriented vesicles. The procedure described by Hokin and Dixon (1979) gave variable results in our hands, perhaps because of difficulties in controlling the amounts of "residual ether" required. On the other hand, freeze–thaw sonication in the presence of optimal cholate (Karlish and Pick, 1981) was reproducible. Indeed, removal of cholate was not necessary when the vesicles were sufficiently diluted during assay. An optimal phospholipid/protein weight ratio of about 45/1 was reported for both freeze–thaw sonication procedures. Hokin and Dixon (1979) noticed a dramatic effect of Na^+ concentration with an optimal concentration of about 20 mM. Na^+ transport was abolished at Na^+ concentrations above 90 mM NaCl. The authors concluded that the effect of Na^+ was on the reconstitution procedure per se, because after reconstitution in the absence of Na^+ no inhibition by addition of high Na^+ during the assay was observed. This conclusion is likely to be correct, since ATP-dependent rates exceeding 200 nmol $\times$ min^{-1} $\times$ mg protein^{-1} were observed with vesicles loaded with 200 mM NaCl when reconstitution was achieved by the cholate freeze–thaw sonication procedure (Karlish and Pick, 1981). Moreover, rather unexpected specific effects of monovalent cations have been observed in the formation of large liposomes, with K^+ being much superior to Na^+ (Oku and MacDonald, 1983). On the other hand, it should be pointed out that inhibition can be caused by the presence of high intravesicular Na^+ concentrations that may impair the efflux of Rb^+ (cf. Skou, 1982).

I have elaborated on this example of apparent controversy to reemphasize the difficulties encountered in separating the parameters that influence reconstitution per se from direct effects on the rate of flux. However, it is also possible that in this particular case some of the differences observed in the two laboratories may be related to the use of ATPase preparations from electric eel and from pig kidney, respectively.

An interesting effect of prolonged sonication was observed with reconstituted Na^+,K^+-ATPase (Hokin and Dixon, 1979). Although 85% of the ATPase activity was sensitive to ouabain following freezing and thawing, after 3 min sonication most of the ATPase activity was lost as well as some transport capacity. These data suggest that unincorporated ATPase was more sensitive to sonic oscillation than incorporated ATPase. Accordingly, sonication for 3 min reduced the Na^+ flux by 50%, whereas 80% of the ouabain-sensitive ATPase was eliminated. It may be possible to exploit this phenomenon by exploring the phospholipid specificity of protection against sonication. An alternative explanation for these findings is that prolonged sonication favors incorporation of the enzyme in the right side out configuration, thereby preventing access of ATP to the active site. This possibility could be tested by including the ATP–ADP transporter during reconstitution and by measuring ATP-driven Rb^+ influx, thereby establishing the presence of right side out vesicles.

The ratio of ATP-driven Na^+ influx to Rb^+ efflux was 1.5 in the vesicles reconstituted by either of the two freeze–thaw reconstitution procedures. A regeneration system for intravesicular K^+ was provided by using valinomycin as an electrogenic K^+ carrier and a proton ionophore to collapse the membrane potential (Hokin and Dixon, 1979). It would be of interest to learn whether Na^+ influx in such a system remains linear for longer than 30 sec, as is apparently the case in the absence of a K^+-regenerating system.

2. Cholate Dialysis

A modified cholate dialysis procedure for the reconstitution of dog kidney ATPase has been described (Goldin, 1979). Excellent data on the stoichiometry of $3:2:1$ for Na^+/K^+-pumped/ATP hydrolyzed were recorded. Based on the virtual lack of ouabain-sensitive ATPase, it was concluded that "an insignificant amount of unreconstituted ATPase activity is present." These vesicles should indeed be considered an ideal reconstituted preparation of the Na$^+$,K$^+$-ATPase were it not for the fact that the rate of Na$^+$ flux was less than 0.4 μmol $\times$ min^{-1} $\times$ mg protein^{-1} compared to 1 to 2 μmol by freeze–thaw sonication and 35 observed with inside out vesicles prepared by a C$_{12}$E$_8$ removal procedure (Cornelius and Skou, 1984). Even correcting for the presence of 50% right side out vesicles (Goldin, 1979), we are left with a large discrepancy in the rate of Na$^+$ flux. Further attempts to improve the cholate dialysis procedure, particularly with a mixture of pure phospholipids and perhaps inclusion of the missing proteolipid, may clarify the discrepancies in the observed rates of flux.

Low rates of Na$^+$ flux were also obtained with pig kidney ATPase using the cholate dialysis procedure of reconstitution (Jørgensen and Anner, 1979). Alterations in the ratio of Na$^+$/K$^+$ flux seen with trypsin-treated Na$^+$,K$^+$-ATPase using the dialysis procedure were not observed when the somewhat more efficient cholate freeze–thaw sonication procedure was used (Karlish and Pick, 1981).

3. Octylglucoside Dilution

There are, as pointed out earlier, several advantages associated with the octylglucoside reconstitution procedure. In the case of the Na$^+$,K$^+$-ATPase from electric eel, rather high concentrations of octylglucoside (2%) were required for maximal rates of Na$^+$ flux (Racker *et al.*, 1979). It should be reemphasized that optimal conditions must be established with each enzyme preparation. We have noted that the Na$^+$,K$^+$-ATPase from Ehrlich ascites tumor cells is much more sensitive to octylglucoside than the electric eel ATPase, and a reliable reconstitution procedure for the tumor enzyme has not as yet been devised.

4. C$_{12}$E$_8$ Removal

A significant advance in the reconstitution of the Na$^+$,K$^+$-ATPase has been reported recently (Cornelius and Skou, 1984). The enzyme from rectal glands of the spiny dog fish was solubilized with C$_{12}$E$_8$ added to sonicated liposomes and freed of C$_{12}$E$_8$ with Bio-Beads. A careful analysis showed that the proportion of inside out vesicles increased from 10% to 30% when the lipid/protein ratio was increased from 10 to 75. Optimal reconstitution was obtained when both phosphatidylinositol and cholesterol were included in the lipid composition. Maximal rates of Na^+ uptake obtained were much higher than reported previously (up to 35 μmol Na^+ per mg protein per min). These high rates were partly due to the mode of calculating activity, in terms of mg protein of inside-out-oriented vesicles, which is indeed the proper way of expressing specific ATP-driven Na^+ flux.

II. The Ca^{2+} Pump and Ca^{2+}-ATPase

A. Physiological Function

Ca^{2+} has profound effects on many enzymes. Some are activated; some inhibited. The intracellular concentration of Ca^{2+} is very low (ca. 0.1 μM), the extracellular concentration approaches 2 mM. Although phospholipid bilayers are very impermeable to Ca^{2+}, there are several transport systems in the plasma membrane that permit its entry and exit, as reviewed elsewhere (Racker, 1980). In excitable cells the addition of, e.g., acetylcholine temporarily opens a flood gate for Ca^{2+}. In cells that contain a large number of coated pits in the plasma membrane, the remarkably rapid recycling of these structures should result in considerable entry of Ca^{2+} by endocytosis. It is not clear how such trapped Ca^{2+} moves inside of the cell, but it obviously must somehow get out of the cell.

It is now apparent (Carafoli and Zurini, 1982) that both excitable and some nonexcitable cells have at least two pathways of Ca^{2+} extrusion: the ATP-driven Ca^{2+} pump and the Na^+–Ca^{2+} exchange transporter. Excellent reviews on the Ca^+-ATPase of plasma membranes have been written (e.g., Schatzmann, 1982), and I shall mainly deal with those of its properties that have emerged from reconstitution studies.

B. Multiple ATPases in Plasma Membranes

The first question is whether there is more than one type of ATP-driven Ca^{2+} pump in the plasma membrane. There are many descriptions of Ca^{2+}-ATPases in plasma membranes that suggest a multitude of different enzymes. However, there is need for caution in the interpretation of these findings. In addition to the problem of eliminating contamination by other Ca^{2+}-activated ATPases, e.g., from sarcoplasmic reticulum and mitochondria (F_1), it is known that the plasma membranes contain an ecto ATPase which is stimulated by addition of Ca^{2+}. The function of this ecto ATPase is unknown (cf. Banerjee, 1981). Thus, evidence for a Ca^{2+}-ATPase involved in Ca^{2+} excretion must depend on measurements of Ca^{2+} transport in either native vesicles derived from the plasma membrane or in proteoliposomes reconstituted with plasma membrane proteins. It appears that there are at least two types of ATP-driven Ca^{2+} pumps in plasma membranes based on such studies. One is present in, e.g., erythrocytes and is activated by calmodulin (Schatzmann, 1982); this enzyme has been reconstituted into liposomes (Haaker and Racker, 1979; Carafoli and Zurini, 1982). The second type is present in plasma membranes of Ehrlich ascites tumor cells (Hinnen *et al.*, 1979). It was reconstituted into liposomes and shown to be independent of added calmodulin. Moreover, conventional calmodulin inhibitors at reasonable concentrations had no effect on Ca^{2+} transport.

The Ca^{2+}-ATPase of plasma membranes responsible for Ca^{2+} pumping is like all other E_1E_2-ATPases sensitive to vanadate (Carafoli and Zurini, 1982). Reported exceptions to this rule should be interpreted with caution, in view of the possibility that the active site of the enzyme may be protected by Ca^{2+}, as shown in the case of sarcoplasmic reticulum ATPase (O'Neal *et al.*, 1979).

C. Purification, Lipid Dependency, and Reconstitution

A major advance in the purification of the calmodulin-stimulated Ca^{2+}-ATPase was the use of calmodulin-affinity chromatography (Carafoli and Zurini, 1982). As observed with many other membrane proteins (Epstein and Racker, 1978), the reconstitutively active enzyme must be protected by phospholipids during purification. The purified enzyme from human erythrocytes contains one major band, as judged by SDS–PAGE analysis (Carafoli and Zurini, 1982), but the Ca^{2+} transport activity after reconstitution appears to be considerably lower than the previously reconstituted Ca^{2+} pump from pig erythrocytes which also traveled as a single major band in SDS–PAGE (Haaker and Racker, 1979). We do not know the reason for this discrepancy, but a high specific activity of biological function is as important or more important than demonstration of a "major" band in SDS–PAGE for the evaluation of the nativity of a pure protein.

Lesson 22:   The virtues and vices of SDS–PAGE

The frequent use of SDS–PAGE becomes apparent to anyone who turns the pages of a biochemical journal of today. Combined with autoradiography, it has become one of the most useful tools in protein biochemistry. But there are problems. Staining with Coomassie blue tells us where the major bands are but is well-known to be inaccurate for quantitative protein evaluations. SDS–PAGE analysis may be treacherous for evaluating purity, as pointed out in Lessons 16 and 17. A "major" band may be representative of a mixture of native and inactivated enzyme or may even hide a protein sister under the skin (a contaminant with the same M_r). SDS–PAGE, moreover, may give inaccurate molecular weights. Perhaps we should use a new symbol, M_{SDS}, suggested by Bernard Davis, when we use SDS–PAGE to characterize a protein. Sugar-coated, phosphorylated, or hydrophobic proteins have different mobilities relative to their true molecular weight when compared with the hydrophilic proteins commonly used as standards. Even dansylated proteins, conveniently used as standards in autoradiography, may have slightly altered mobilities.

Interesting differences in Ca^{2+}-ATPase activity have been observed in the presence of PC (Gietzen *et al.,* 1980) or PS (Carafoli and Zurini, 1982). With PC, the enzyme was stimulated by calmodulin; with PS, maximal activity was seen in the absence of calmodulin and there was a change in K_m for Ca^{2+}. Stimulation of ATPase activity was noted in the presence of Tween 20 which markedly inhibited Ca^{2+} uptake in reconstituted vesicles (Haaker and Racker, 1979).

A particularly interesting observation is the stimulation of the Ca^{2+}-ATPase from plasma membranes of rat brain synaptosomes by low concentrations (2 μM) of phosphatidylinositol 4,5-bisphosphate (Penniston, 1982). Without data from reconstitution experiments it is difficult, however, to evaluate this phenomenon. Is this phospholipid the physiological equivalent of Tween 20, acting as an uncoupler of the Ca^{2+} pump, or is it capable of stimulating Ca^{2+} transport as well as ATPase activity like calmodulin? Could this phospholipid play a role in regulating the second type of Ca^{2+}-ATPase that does not respond to calmodulin? These are interesting question, particularly in relationship to the "phosphoinositol response" of plasma membranes (Mitchell *et al.*, 1981); they can be answered by reconstitution experiments. Moreover, such experiments may also shed light on the interaction of proteins with specific phospholipids and on the role of the remarkably complex lipid composition of membranes in general, an important subject about which we know next to nothing.

The purified Ca^{2+}-ATPase activity of pig erythrocytes after reconstitution into liposomes was stimulated about threefold by A-23187 (Haaker and Racker, 1979) similar to the stimulation observed with the sarcoplasmic reticulum Ca^{2+}-ATPase (Banerjee *et al.*, 1979). A tenfold stimulation of ATPase activity by A-23187 was observed with the reconstituted enzyme from human erythrocytes (Niggli *et al.*, 1982). These authors observed "extra" protons in the medium (in excess of those produced by hydrolysis of ATP) and propose that the Ca^{2+} pump of plasma membranes catalyzes an electroneutral Ca^{2+}–H^+ exchange. Others (Waismann *et al.*, 1981; Smallwood *et al.*, 1983) have considered an electrogenic mode of action of this pump. Perhaps such apparent contradictions are more semantic than real. As will be illustrated in the case of the Ca^{2+} pump of sarcoplasmic reticulum, for all practical purposes the transport of Ca^{2+} in SR is electroneutral. Yet, it is possible under special conditions to demonstrate that it is basically electrogenic (Zimniak and Racker, 1978). A similar situation has evolved in the case of the K^+,H^+ pump of the stomach, which for many years has been referred to as an electroneutral pump. Yet, under some conditions it was shown to operate as an electrogenic pump with Cl^- as co-ion (Faller *et al.*, 1982). Once we have penetrated deeper into the mechanism of action of E_1E_2 pumps we shall probably find that they all operate electrogenically. If the mechanism of counterion transport is indeed a sequential one, it is mandatory that, however fleetingly, the process is electrogenic. Reconstitution experiments should be most helpful in probing into this question, although the possibility must be considered that in the course of purification, the ATPase protein may be altered, inducing a leakiness that obscures the electrogenic mechanism. We thus are reminded of the necessity to remain aware of the physiology and its haunting complexity.

Lesson 23: On the difference between physiology and biochemistry

As a biochemist I have stressed the need for obtaining pure proteins more than once. I have also stressed the need to distrust data obtained with crude systems. As a physiologist I want to stress the need to pay attention to data obtained with native membranes or even with intact cells and to distrust data obtained with pure proteins. The opposing views of physiologists and biochemists have been healthy and have fertilized progress. What we must fear is the day they start agreeing. The poet Walt Whitman said: "Do I contradict myself? Very well, then I contradict myself."

A Ca^{2+}-ATPase has been isolated from heart sarcolemma and reconstituted into crude soybean liposomes by the cholate dialysis procedure (Caroni *et al.*, 1982). The enzyme is similar to the erythrocyte ATPase with respect to M_r, K_m for Ca^{2+} and ATP, sensitivity to vanadate, and activation by calmodulin, acidic phospholipids, or trypsin. In view of the fact that this enzyme was isolated from excitable cells, it is of particular interest to note that both Ca^{2+} translocation and ATPase activity appear to be regulated in sarcolemmal vesicles by phosphorylation–dephosphorylation reactions. Phosphorylase phosphatase inhibited the ATPase and a protein kinase present in the vesicles reversed the inhibition induced by the phosphatase. Thus far, however, similar results could not be achieved with the reconstituted purified enzyme. Another interesting example, emphasizing the importance of performing comparative biochemistry, is the regulation of a plasma membrane Ca^{2+}-ATPase of adipocytes by insulin (McDonald *et al.*, 1982). Insulin inhibits ATPase activity and modulates calmodulin binding. Unfortunately, its effect on transport is more difficult to measure in inside out vesicles that are impermeable to insulin or right side out vesicles that are impermeable to ATP and do not transport Ca^{2+}. It should be possible by reconstitution experiments to overcome these obstacles and to explore this and some of the other curious observations, such as differences in Mg^{2+} requirements for ATPase activity and Ca^{2+} transport.

It is apparent that much more work is needed to be done on the plasma membrane Ca^{2+} pumps with intact cells, in native vesicles, and in reconstituted systems. If each Ca^{2+} pumped out results in the import of two H^+ into the cell, how are these protons disposed of? If this disposal is linked to the Na^+/H^+ antiporter and secondarily to the Na^+,K^+ pump, interesting experiments could be designed with ouabain and amiloride or more potent analogs that inhibit $Na^+–H^+$ exchange; or it could be linked to an ATP-driven proton pump that we shall discuss later. On the other hand, if the pump is electrogenic, which ions are responsible for collapsing the membrane potential?

III. The H$^+$,K$^+$-ATPase of the Gastric Mucosa

The parietal cells of the gastric mucosa, involved in gastric excretion, contain an ATP-driven H$^+$ pump with K$^+$ as a counterion (Ganser and Forte, 1973). Of special interest is the role of this pump in the secretion of HCl in the stomach, a puzzling process resulting in the generation of an H$^+$ gradient of $10^{6.6}$ (Sachs et al., 1979). Its mode of action resembles that of the Na$^+$,K$^+$-ATPase. There is little doubt that the transformer is a 100,000-dalton protein, but the homogeneity of the purified enzyme is controversial. SDS–PAGE yielded a sharp single band corresponding to 100,000 daltons, whereas electrofucusing of the same preparation revealed several bands (Sachs et al., 1979). Based on these and other data obtained by digestion with trypsin, Saccomani et al. (1979) concluded that three proteins were hiding behind the 100,000-dalton band: a catalytic subunit containing the ATP binding site, a glycoprotein, and an unknown third protein, not protected against trypsin by ATP. The proposition of a functional trimer is consistent with an estimate of molecular mass of about 300,000, based on radiation inactivation studies by Saccomani et al. (1981).

On the other hand, Peters et al. (1982) analyzed a preparation of H$^+$,K$^+$-ATPase, purified by an alternative procedure that included gel filtration, and could not find evidence for heterogeneity by four different methods of analysis (which, however, did not include electrofocusing). The resemblance with respect to amino acid and carbohydrate composition between this preparation and the α subunit of Na$^+$,K$^+$-ATPase is indeed striking and was proposed to be evidence against the presence of other subunits. Thus, the question of subunit composition of this enzyme must be considered still unsettled. Moreover, a decisive answer to the subunit composition of the physiologically active transporter will not emerge until reconstitution experiments with the purified enzyme and/or with its proposed subunits become feasible. Once again we are faced with the same difficulties encountered with the subunits of the Na$^+$,K$^+$-ATPase.

Digestion with phospholipase A$_2$ of vesicles capable of H$^+$/K$^+$ transport resulted in loss of ATPase activity that could be restored by addition of phosphatidylethanolamine. Other phospholipids were less effective (Sachs *et al.*, 1979).

A significant development is the demonstration of electrogenicity (Faller *et al.*, 1982) which I mentioned before. It may have an important bearing on the physiological process of HCl secretion. There are many observations that suggest that the H^+,K^+-ATPase is a participant in HCl excretion. However, many inhibitors of HCl excretion (particularly thiocyanate) do not affect the H^+,K^+ pump. Moreover, it cannot account for the enormous physiological proton gradient. It also leaves unexplained the absolute requirement for oxygen in acid secretion. Thus it appears that more than one pump may be involved in acid excretion. A possible candidate is the recently discovered proton pump in plasma membranes (Stone *et al.*, 1984a), resembling that of clathrin-coated vesicles, which is sensitive to NEM and compulsorily linked to a chloride flux (Xie *et al.*, 1983). It could explain the previously recorded ATP-driven uptake of Cl^- in vesicles obtained from rabbit stomach (Soumarmon and Racker, 1978). In the case of clathrin-coated vesicles, a similar ATP-driven Cl^- uptake was shown to be secondary to the electrogenic ATP-driven uptake of protons (Xie *et al.*, 1983). This formulation would account for the old observation that proton excretion by gastric mucosa is very low in chloride-free medium (Rehm, 1965). It is also consistent with the sensitivity of ATP-driven Cl^- uptake in gastric vesicles to NEM and resistance to vanadate, but it is inconsistent with the lack of sensitivity to DCCD (Soumarmon and Racker, 1978).

In any case, where are all the protons coming from, and how can we explain the oxygen requirement? One candidate is H_2O in the presence of respiratory CO_2 and carbonic anhydrase. An alternative source of protons may be the substrates of oxidative phosphorylation (e.g., succinate or fatty acids). Normally the proton gradient is utilized for the generation of ATP. However, if the flux of protons is incompletely coupled to ATP generation because of either a defective or partly defective, slipping proton pump or an altered membrane, the protons are excreted by the mitochondria and become available in the cytoplasm for export together with Cl^-. A good model for such a loosely coupled system is encountered in submitochondrial particles prepared in the absence of divalent cations or in mitochondria that have been stored frozen (Racker and Horstman, 1972). They respire in the absence of ADP and P_i, yet are capable of generating ATP in the presence of ADP and P$_i$. A more physiological example for a "loosely coupled" electron transport chain are thylakoids that reduce NADP in the absence of ADP. A pathological example is that of loosely coupled mitochondria in humans afflicted with a muscle disease (Luft *et al.*, 1962). Operation of such a loosely coupled respiratory chain in parietal cells of the stomach seems particularly attractive. It could provide some ATP in addition to that generated by glycolysis, explaining some observations made by Soumarmon and Racker (1978), yet provide an H$^+$-generating system that can be used for the export of HCl.

Lecture 7

ATP-Driven Ion Pumps in Organelles, Microorganisms, and Plants

> The difficulty lies,
> not in new ideas,
> but in escaping the
> old ones.
>
> **J. M. Keynes**

I. The Ca^{2+} Pump of Sarcoplasmic Reticulum and Related Organelles

The Ca^{2+}-ATPase of sarcoplasmic reticulum deserves special consideration for several reasons. It plays a pivotal role in the regulation of intracellular Ca^{2+} and in the muscular contraction–relaxation cycle. It can be readily purified (MacLennan, 1970) and was the first among the E$_1$E$_2$ ATPases to be reconstituted (Racker, 1972b). It can be reconstituted by a variety of methods with well-defined lipids that influence its mode of operation and efficiency (Caffrey and Feigenson, 1981; Navarro *et al.*, 1984).

There are numerous reviews on the Ca^{2+}-ATPase pump (MacLennan and Holland, 1975; Racker, 1978; deMeis, 1982; Hasselbach and Waas, 1982) and once again I shall emphasize only those aspects that have emerged from resolution-reconstitution studies.

A. The Latency of the Ca^{2+}-ATPase in SR Vesicles

The enzyme has been purified from SR vesicles about tenfold from a specific activity of 3 to about 30 μmol $\times$ min^{-1} $\times$ mg protein^{-1} (MacLennan, 1970), yet the estimates of the enzyme content in SR ranges from 60–90% of the total protein present in SR vesicles (MacLennan and Holland, 1975; Fleischer and McIntyre, 1982). This discrepancy is partly accounted for by inhibition of ATPase activity by Ca^{2+} accumulated inside the vesicles, since A23187 stimulates ATP hydrolysis in SR vesicles three to fourfold (Banerjee *et al.*, 1979). An additional control of ATP hydrolysis appears to be exerted by the phospholipid composition, as revealed from reconstitution studies (Knowles *et al.*, 1975, 1976; Navarro *et al.*, 1984).

In contrast to vesicles derived from plasma membranes and mitochondria which are often predominantly in the inside out configuration, the vesicles derived from sarcoplasmic reticulum are right side out. This greatly facilitates the exploration of the mode of action of the pump, its reversibility, and partial reactions (Racker, 1978).

B. The Proteins of the Ca^{2+} Pump of SR Vesicles

Most purified preparations of the Ca^{2+}-ATPase consist of two major components, as revealed by SDS–PAGE analysis: a 100,000-dalton protein that interacts with ATP and represents the energy transformer of the pump; and a second component, a proteolipid, characterized in pioneering studies by MacLennan *et al.* (1972). It was extracted with chloroform–methanol and contained covalently attached fatty acids. Early estimates of its molecular weight were probably premature, since multiple proteolipids have been extracted from SR with different amino acid compositions and different minimal molecular weights (Knowles *et al,* 1980). The role of proteolipids in the Ca^{2+} pump is still uncertain. It should be noted that preparations of Ca^{2+}-ATPase thoroughly delipidated and reactivated with C$_{12}$E$_8$ still contain the proteolipid (Dean and Tanford, 1977). Moreover, in preparations lacking the low-molecular-weight band, the possibility of an association between the highly hydrophobic proteolipid and the 100,000-dalton protein that cannot be dissociated by SDS has not as yet been completely eliminated. We are continuing to search for new methods of purifying proteolipids without the use of solvents.

It is now apparent that in reconstituted vesicles the pure 100,000-dalton protein can function as a Ca^{2+} pump (Knowles *et al.*, 1980; MacLennan *et al.*, 1980). On the other hand, both papers report that the efficiency of pumping, as expressed by the ratio of Ca^{2+} transported/ATP hydrolyzed, is low in such reconstituted vesicles (1.0 or below), whereas in SR vesicles and in reconstituted vesicles with ATPase preparations that contain the proteolipid, the ratio approaches 1.7 (Racker and Eytan, 1975). It is not clear whether the proteolipids have a direct role in the function of the pump or whether they are involved in the organization of the pump during reconstitution and perhaps also during assembly *in vivo*. Unfortunately, proteolipids are notoriously sensitive to chloroform–methanol, even in the case of H^+ pumps of mitochondria and bacteria, where their function has been clearly established. More of this later.

The purified enzyme catalyzes the hydrolysis of ATP that is dependent on both Mg^{2+} and Ca^{2+}. The purified enzyme also catalyzes a $^{32}P_i$–ATP exchange (Racker, 1972b) as well as the stoichiometric formation of net ATP from P_i and ATP in a two-step reaction sequence that is dependent on Mg^{2+} in the first step and on Ca^{2+} in the second (Knowles and Racker, 1975a). The presence of Ca^{2+} during the first step prevents the overall reaction of ATP formation. Based on these observations a mechanism of ATP formation has been proposed and extended to the mechanism of ATP-driven Ca^{2+} transport (Racker, 1977b). The most important and novel feature of this model is that the binding of Mg^{2+} to the enzyme is proposed to induce a conformational change in the protein that allows the formation of the phosphorylated intermediate, $E_2 — P$, from inorganic phosphate. The subsequent steps are conformational changes of the protein, allowing the formation of $E_1 \sim P$ and phosphotransfer to ADP to form ATP. In reverse, in the presence of ATP the same conformational changes drive Ca^{2+} into the vesicles in a continuous process. This formulation (schematically presented in Fig. 7-1) differs fundamen-

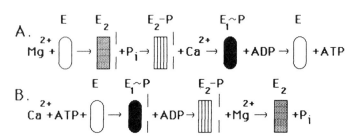

Fig. 7-1 The Mg^{2+}-cycle hypothesis. (A) ATP formation without ion gradient; (B) Continuous Ca^{2+} transport and ATP hydrolysis.

tally from other proposals because, firstly, it assigns a specific function to Mg^{2+} as inducer of a conformational change; secondly, it assigns a specific function to Ca^{2+} as a displacer of Mg^{2+}; and, thirdly, it requires cyclic opening and closing of the channel. Indeed, the major function of the membrane is visualized to regulate the opening and closing of the channel. This facilitates the on and off interactions of the enzyme with Mg^{2+} that induce the cyclic conformational changes of the protein. In the formation of ATP, Ca^{2+} from inside of the vesicles is used to displace Mg^{2+} from the active site into the medium. In the transport of Ca^{2+}, the ATP energy is used to allow the displacement, by external Mg^{2+} of Ca^{2+} from the active site into the vesicular compartment.

It may be a consolation to some young (and old) scientists that we had great difficulties in getting the paper on net ATP formation without a Ca^{2+} gradient accepted for publication. I still consider it among the most significant contributions from my laboratory.

Lesson 24: On the character of scientists

Scientists can be as illogical as any other mortals. How else can we account for statements that keep appearing in our most prestigious journals such as "The chemicals used were of highest purity available." Do the editors really believe that the investigators have compared the chemicals from dozens of available sources?

On the other hand most scientists are by nature skeptical when confronted with new concepts (proposed by other scientists). Thus, papers with novel concepts, e.g., the Krebs cycle and several others (that modesty does not allow me to identify), were first rejected for publication, while dull and conservative papers from the same laboratory (that pride does not allow me to identify) get readily accepted.

Among the major unresolved questions is the exact relation between the Ca^{2+} channel and the ATP transformer, both part of the 100,000 molecular weight protein. Important advances have been made by proteolytic fragmentation of the protein and by sequence analyses (MacLennan *et al.*, 1980; Green *et al.*, 1980). Cleavage with trypsin revealed in SDS–PAGE two fragments (A, 60,000 daltons; B, 55,000 daltons). On longer digestion, A was cleaved to A_1 (33,000 daltons) and A_2 (25,000 daltons.) It is of interest that these cleavages, which can be seen only after treatment with SDS, do not impair the ATPase activity. A_2 has been fragmented with cyanogen bromide, and one of the resulting peptides (13,000 daltons) exhibited Ca^{2+}-dependent ionophoric activity in black lipid membranes (MacLennan et al., 1980). The 25,000-dalton A_2 fragment also contained the site that interacts with DCCD and that was protected against DCCD inactivation by low concentrations of Ca^{2+} (Pick and Racker, 1979b). Based on studies of the trypsin fragments, MacLennan et al. (1980) have formulated a tentative diagram for the folding of the protein within the membrane, resembling in complexity that of bacteriorhodopsin.

C. The Phospholipids of the Ca^{2+}-ATPase Pump of SR Vesicles

The phospholipid composition of the purified Ca^{2+}-ATPase is similar to that of intact SR (MacLennan and Holland, 1975). The major phospholipid is phosphatidylcholine (65%); other constituents are phosphatidylethanolamine (18%), phosphatidylinositol (9%), phosphatidylserine (4%), and several other minor lipids. The protein has been delipidated to about 4–5 moles of residual phospholipid per mole of enzyme, resulting in the reversible inactivation of the ATPase (Knowles et al., 1976; Dean and Tanford, 1977). Pure phosphatidylcholine or even dodecyloctaoxyethylene-glycol monoether $(C_{12}E_8)$ restored ATPase activity to the delipidated enzyme (Dean and Tanford, 1978). The role of the phospholipids becomes more intricate when the individual steps of the reaction are examined (Knowles et al., 1976). Considerably less lipid was required for the restoration of phosphoenzyme formation with [γ-^{32}P]ATP than for ATPase activity. N-Acetyl dilaurylphosphatidylethanolamine was at least as effective as phosphatidylcholine in restoring $E_1{\sim}P$ formation but was less than 5% as effective in restoring ATPase activity. Since even a simple detergent such as $C_{12}E_8$ can activate the hydrolysis of ATP, it is possible that the acetyl group interferes with the hydrolysis of $E_2 — P$ or with the formation of $E_2 — P$ from $E_1{\sim}P$. This possibility was supported by the observation that addition of N-acetyl dilaurylphosphatidylethanolamine to native ATPase suppressed ATP hydrolysis at concentrations that had no effect on $E_1{\sim}P$ formation. These clues become important in the interpretation of the data on reconstitution of Ca^{2+} transport with N-acetyl phosphatidylethanolamine, that will be discussed later.

An interesting controversial issue is the concept of "boundary lipids." About 30 moles of phospholipid per mole of Ca^{2+}-ATPase remain attached to the protein after extraction with cholate. It was proposed that these lipids form an immobilized "annulus" around the protein, thus representing a case of specific phospholipid–protein interaction (Metcalfe and Warren, 1977). The annulus concept has been recently challenged by several investigators who came to the conclusion, based on studied with proton, deuterium, and phosphorus NMR (Seelig *et al.,* 1981; Fleischer and McIntyre, 1982), that there does not exist an immobilized layer of boundary lipids. On the other hand it is apparent (Watts, 1981) that spectroscopists still do not agree mainly because the magnitude of motional inhibition depends on the method used to detect it. Thus NMR is not a sensitive probe; the time resolution of EPR is orders of magnitude greater. There are examples, such as cytochrome oxidase and Ca^{2+}-ATPase, in which a few molecules of phospholipids are associated more firmly with the protein than the bulk lipids. We should therefore probably not abandon the concept of boundary lipids but use it in a restricted sense. Even the NMR observations (Fleischer and McIntyre, 1982) that "the protein introduces only a minor perturbation on the motion of the phospholipid" is consistent with the proposition that a few molecules of phospholipids have a firm association with the protein and are difficult to remove without its denaturation. Although the original concept of the annulus is no longer viable in view of the demonstrated mobility of these lipids (Seelig *et al.,* 1981), specific associations between proteins and lipids that escaped detection by spectroscopists need further biochemical exploration. This should be particularly feasible in reconstituted systems that allow measurements of biological function. It was pointed out (Watts, 1981) that such special lipids may be important: (a) for efficient sealing of the membrane at the site of insertion, e.g., of a transport protein; (b) to provide a motional buffer between the protein and the bulk lipids; (c) to sterically maintain at equilibrium the active conformation of a membrane-bound enzyme, which would require a fast motional interaction on the time scale defined by the catalytic turnover rate of 10^2–10^3 per second; and (d) to contort and conform to the protein interface to solvate it in a way that minimizes the energy configuration within the whole membrane. The two latter points are particularly relevant with respect to the previously mentioned need to regulate the opening and closing of a channel that must be associated with the transport protein and that allows the ion translocation across the membrane.

D. Reconstitution of the Ca^{2+} Pump of SR Vesicles and the Role of Lipids

Several widely used methods are available for the reconstitution of the Ca^{2+} pump of sarcoplasmic reticulum; cholate–deoxycholate dialysis (Knowles and Racker, 1975b), freeze–thaw sonication (Zimniak and Racker, 1978), and octylglucoside dilution (Racker *et al.*, 1979). A variant of the detergent dialysis method is a procedure that was developed by Fleischer and his collaborators (cf. Wang *et al.*, 1979). It consists of solubilizing SR membranes with deoxycholate and restoring pump activity with its own phospholipids. The method, which I call resurrection, is different from that of reconstitution, which aims to restore activity with the minimal requirement of individual components. However, it has yielded interesting data that deserve further exploration. Perhaps the most important conclusion is that it is possible to recover full activity after solubilization of the membrane with SR lipids, forming vesicles with a low phospholipid-to-protein ratio (0.38 μmol lipid/mg protein). The second interesting observation is that asymmetry was lost, which is in contrast to observations on vesicles reconstituted with excess phospholipids (Knowles and Racker, 1975b).

Although I have discussed earlier some of the problems that have emerged from the use of these different methods, it seems appropriate to return to them with specific reference to the question of firmly bound lipids and their possible role in the apparently discrepant lipid requirements, depending on the method of reconstitution. Reconstitution by cholate–deoxycholate dialysis (Knowles and Racker, 1975b) revealed that phosphatidylethanolamine is required for Ca^{2+} transport (Knowles *et al.*, 1975). Reconstitution with *N*-acetyl phosphatidylethanolamine yielded vesicles that lacked both ATPase and Ca^{2+} transport activities that were restored by addition of small amounts of hydrophobic amines (e.g., stearoylamine). Hidalgo *et al.* (1982) demonstrated that blockage of the amino group of phosphatidylethanolamine in sarcoplasmic reticulum vesicles with fluorescamine inhibited Ca^{2+} transport without affecting ATPase activity. These observations are consistent with the proposal that the amino group of phosphatidylethanolamine is essential for Ca^{2+} transport, but suggest also that they are not essential for ATP hydrolysis. The above mentioned inhibition of ATPase by phosphatidylethanolamine acetylated on the amino group may well be caused by a replacement of endogenous phosphatidylethanolamine firmly associated with the enzyme.

This brings me to the data obtained with the Ca^{2+} pump reconstituted by freeze–thaw sonication (Zimniak and Racker, 1978; Caffrey and Feigenson, 1981). In contrast to vesicles reconstituted by the detergent dialysis procedure, vesicles reconstituted by sonication with only phosphatidylcholine catalyze ATP-driven Ca^{2+} transport (though not nearly as well as phosphatidylcholine–phosphatidylethanolamine vesicles). This discrepancy between the two reconstitution methods suggests that either the dependency for phosphatidylethanolamine is not for function but for reconstitution, or that the enzyme reconstituted without detergent contains enough tightly bound PE to satisfy the requirement for amino groups. Reconstitution experiments with delipidated enzyme or even with an enzyme in which PE is replaced by acetyl PE should shed further light on this puzzling problem.

In order to explore further the role of PE, we considered the possibility that coupling of the Ca^{2+}-ATPase depends on the presence of lipids capable of forming nonbilayer structures. It is known from the work of Cullis et al. (1983) that under certain conditions PE tends to assemble into hexagonal II structures. We have reconstituted Ca^{2+}-ATPase by a procedure that allows the use of relatively low lipid-to-protein ratios (2–5 mg lipid/mg protein) and conducted parallel examinations of Ca^{2+} transport and of the tendency of the various lipids to form hexagonal II structures (Navarro et al., 1984). With either dioleoyl PE or monogalactosyl diglyceride, which have a tendency to form such structures, we observed high initial rates of Ca^{2+} transport and high coupling ratios (Ca^{2+} transported/ATP hydroylzed) up to 1.2. Ca^{2+}-ATPase reconstituted with lipids that contain either an increased number of methyl groups or two rather than one galactosyl groups revealed a decrease in both rate and efficiency (Ca^{2+}/ATP ratio) of transport. We could detect hexagonal II structures in freeze-fracture electron micrographs of vesicles reconstituted with either monogalactosyl diglyceride or dioleoyl PE and dioleoyl PC at a molar ratio of 12:1 and a lipid-to-protein ratio of 3:1, and kept at 37°C. Based on comparative studies with several phospholipid mixtures, we proposed that cone-shaped lipids capable of hexagonal II structure formation, such as dioleoyl PE and monogalactosyl diglyceride, yield reconstituted vesicles that catalyze rapid Ca^{2+} transport with high efficiency. On the other hand, we must recognize that native sarcoplasmic reticulum vesicles do not have a phospholipid composition that resembles that of these reconstituted vesicles. We must therefore assume that either an unidentified lipid, protein, or proteolipid, as proposed earlier (Racker and Eytan, 1975), contributes to the high efficiency of native vesicles.

I would like to generalize by suggesting that the remarkably high efficiency of all ATP-driven pumps in normal cells is achieved by sheltering the active site against the illicit entry of water. In the case of the mitochondrial pump that is exposed to an aqueous environment, this is achieved by a regulatory polypeptide. This subunit is a mitochondrial ATPase inhibitor (Pullman and Monroy, 1963) that does not interfere with ATP generation. As just suggested in the case of the lipid-embedded ATP-driven Ca^{2+} pump of SR, a specific lipid, protein, or proteolipid fulfills the function of preventing entry of water and slipping of the machinery (Fig. 7-2). An alternative explanation has been proposed by Pickart and Jencks (1984). Instead of protecting the active site against water entry, it is visualized that Ca^{2+} is occluded under conditions of coupling and is channeled into the interior of the vesicles during efficient pumping. When the pump is impaired, Ca^{2+} is not occluded and released to the outside. This appealing hypothesis may in fact be more suitable for experimental verification than our proposal. It would be particularly interesting if the proposed occluded Ca^{2+} operating during coupling is the

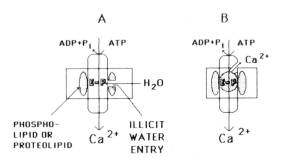

Fig. **7-2** Slippage of Ca^{2+} pump. (A) Hydrolysis of $E_1{\sim}P$; (B) Ca^{2+} recycling.

same as the occluded Ca^{2+} that protects the enzyme against inhibition by DCCD (Pick and Racker, 1979b). Similarly, one could postulate shielding of occluded H^+ or K^+ (Karlish and Stein, 1982) in the case of the proton and sodium pumps.

It was recently observed that the Na^+,K^+ pump can be rendered inefficient by small amounts of neutral detergents (Racker and Riegler 1985; Racker, 1985), the Ca^{2+} pump by gentle proteolysis (Scott and Shamoo, 1982) or by small amounts of duramycin (Navarro et al., 1985), and the mitochondrial pump by anaesthetics (Rottenberg, 1983). In all three cases this took place without induction of an ion leak (Fig. 7-3). Thus, it appears that the change occurs at the molecular level of the pump. We refer to it as "loose coupling" or "shunting out" (Lehninger et al., 1959) or as "molecular slipping" (Wikström et al., 1981a,b), but this does not tell us much about the responsible molecular mechanism. Moreover, contrary to implications in the literature, this kind of pathology does not rule out the operation of a basic mechanism of coupling, such as described by the chemiosmotic hypothesis.

Lesson 25: Inefficient, loosely coupled, and slipping pumps

For many years mitochondriologists (often referred to as mitochondriacs) have used the term "loosely coupled" to describe mitochondria or submitochondrial particles that respire at a high rate in the absence of an uncoupler, but generate ATP in the presence of P_i and ADP, albeit inefficiently. I mentioned earlier the case of a loosely coupled women. I could call her more accurately a lady with loosely coupling mitochondria, but I don't think we gain much by calling her a slippery woman or a lady with slipping mitochondria. This is loose and slippery terminology, but what needs to be recognized is that we can either uncouple a pump by making the membrane leaky to ions, or we can make it inefficient by somehow damaging its gears (Fig. 7-3). Sometimes, when there is a partial membrane leak, it is difficult to tell the difference. Once we have made the diagnosis of a defect in the pump itself, we can look for changes in its components. We can try to imitate the defect by rough treatment of a normal pump (postdoctoral modifications), or we can induce genetic and posttranslational modification of the pump components. Emerging data on phospho-

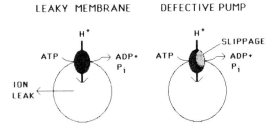

Fig. 7-3 Two mechanisms of pump inefficiency.

rylation and dephosphorylation of pumps and receptors open a new chapter in the regulation and pathology of membrane activities.

By using the freeze–thaw reconstitution procedure with phosphatidylcholine, it was shown that the Ca^{2+} pump is electrogenic (Zimniak and Racker, 1978). This was demonstrated by using valinomycin at different concentrations of K^+ inside and outside the vesicles as a biochemical voltage clamp. Alternatively, it was shown that lipophilic anions accelerate the rate of Ca^{2+} transport in these vesicles.

An interesting study was conducted on the reconstitution of the Ca^{2+} pump by freeze–thaw sonication with phosphatidylcholines of different fatty acid chain lengths (Caffrey and Feigenson, 1981). In fact, as far as I know, this is the only study of this type using a transport assay rather than ATPase activity. The data show that fatty acids with less than 16 carbons did not support a functioning Ca^{2+} pump. One phospholipid was available with fatty acids containing 24 carbons. It was decidedly less effective than the phospholipid with fatty acids containing 16 to 22 carbons. It thus appears that the thickness of the membrane is a determining factor in the operation of the pump.

E. The Ca^{2+} Pump of Cardiac Sarcoplasmic Reticulum

This pump is of special interest because of its regulation by phosphorylation–dephosphorylation mechanisms. The important discovery (Tada *et al.*, 1974; Kirchberger and Tada, 1976) that cAMP-dependent protein kinase stimulates the rate of Ca^{2+} uptake by cardiac SR (without change in efficiency) has opened a new field in the control of ion fluxes. Phosphorylation of phospholambdan, a proteolipid of about 22,000 daltons, by either cAMP- or calmodulin-dependent protein kinase is associated with increased Ca^{2+} uptake, and represents an interesting model for a dual control mechanism (Le Peuch *et al.*, 1979). It should be noted that phospholambdan has been extracted by a mild procedure at a very low concentration of deoxycholate (0.3 μg per mg cardiac SR protein) and purified to homogeneity (Bidlack *et al.*, 1982). Since the Ca^{2+}-ATPase of cardiac SR was also purified by a rather gentle procedure, reconstitution of this control mechanism should be feasible.

F. The Ca²⁺ Pump of Related Organelles

It has long been known (Otsuka *et al.*, 1965) that microsomal preparations of noncontractile cells catalyze an ATP-driven uptake of calcium similar to that catalyzed by SR. There is little known about the resolution and reconstitution of these pumps. Studies similar to those performed with cardiac SR concerning the stimulation by cAMP may be revealing in organelles that are involved in specific physiological functions, e.g., in platelets (cf. Carrol and Cox, 1983) or in pancreatic islet cells (Colca *et al.*, 1982). Ca^{2+} storage in "microsomes" has gained new significance in view of the release of Ca^{2+} by inositol triphosphate (Berridge and Irvine, 1984).

II. ATP-Driven H⁺ Fluxes in Organelles

An ATP-dependent proton flux has been observed in several organelles. In chromaffin granules, lysosomes, synaptosomes, and clathrin-coated vesicles, the basic process appears to be similar, yet distinct differences with respect to anion fluxes and sensitivity to inhibitors have been observed. For example, in clathrin-coated vesicles (Xie *et al.*, 1983) and synaptosomes (Cidon *et al.*, 1983) there is an absolute dependency on Cl^-, whereas in the chromaffin granules Cl^- has a stimulatory effect but is not essential (Cidon *et al.*, 1983). In lysosomes, P_i appears to be the major co-ion (Schneider, 1983).

A. Chromaffin Granules

These organelles, which can be isolated in high yield, have been subjected to extensive analysis for the past two decades. They are capable of accumulating 2.5 μmol of catecholamine per mg of protein which corresponds to an intravesicular concentration of 0.5 M and a catecholamine gradient of greater than 10^4. Only a small part of this enormous gradient of positively charged amine is balanced by negatively charged intravesicular ATP (0.12 M) and acidic proteins (Knoth et al., 1982). It is now generally agreed (Johnson et al., 1982) that an ATP-driven proton pump is responsible for the active monoamine transport. Two or more H^+ move out of the vesicles for each protonated monoamine that moves in.

An H^+ translocating ATPase is responsible for the generation of a $\Delta \bar{\mu}_{H^+}$ which consists of a Δ pH and a membrane potential $\Delta \psi$, similar to the proton pump of mitochondria. There has been, however, some confusion with regard to the H^+-ATPase that is responsible for the generation of the electrochemical proton gradient in the chromaffin vesicles (Apps and Schatz, 1979; Apps, 1982; Cidon and Nelson, 1982). It is now apparent that most preparations of chromaffin vesicles are contaminated with mitochondrial membranes. Thus, claims of ATP synthesis (Roisin et al., 1980) have to be viewed with caution. Mitochondrial F_1 could be removed by treatment with NaBr, and the inactivated membranes were incorporated into liposome vesicles by the cholate dilution procedure, giving rise to an active proton pump (Cidon et al., 1983). The history of the confusion created by mitochondrial contamination is an excellent illustration of the need to establish, by decisive reconstitution experiments, which protein components are associated with biological function. We shall encounter the same problem in other organelles such as the lysosomes that contain more than one ATP hydrolyzing enzyme.

A study of the proton gradient established in chromaffin granules revealed an interesting phenomenon. Addition of ascorbic acid together with phenazine methosulfate resulted in an efflux of protons (Cidon et al., 1983). Neither 2,6-dichloroindophenol nor mercaptoethanol substituted for these two reagents. Moreover, neither lysosomes nor synaptosomes responded in this way. It is therefore possible that this phenomenon may be related to the remarkable ability of chromaffin granules to concentrate ascorbic acid to intravesicular concentrations, as high as 22 mM (Ingebretsen et al., 1980). The mechanism of this transport is still unknown. The transporter responsible for the H^+/catecholamine flux has been solubilized and reconstituted into liposomes (Maron et al., 1979).

B. Lysosomes

Lysosomes are required for the intracellular degradation of proteins. Impairment of their function results in disorders such as lysosome storage diseases (Neufeld *et al.*, 1975). The lysosomal proteinases have an acidic pH optimum and cease to function properly when the intralysosomal pH is high. An increased degradation of proteins by added ATP was first shown in isolated lysosomes by Mego *et al.* (1972). Convincing evidence for an ATP-driven proton pump in lysosomes was presented by Schneider (1981). GTP stimulated less than ATP and nonhydrolyzable analogs of ATP were inactive. DIDS at low μM concentrations inhibited H^+ transport without much effect on ATPase activity. This as well as other observations such as an ATP-driven P_i transport (Schneider, 1983) suggest a resemblance of this cotransport with that occurring in clathrin-coated vesicles (Xie *et al.*, 1983) and in sarcoplasmic reticulum (Carley and Racker, 1982). In all three systems an ATP-driven uptake of anions is facilitated by an ATP-generated positive membrane potential due to the active transport of cations. It was shown by Ohkuma *et al.* (1982) that the lysosomal proton pump is indeed electrogenic. The observation that valinomycin plus K^+ greatly enhanced the effect of FCCP is consistent with an electrogenic mechanism in lysosomes (Cidon *et al.*, 1983). With the disappearance of ion pumps that operate electrogenically, we can now design with greater confidence experiments that have thus far resisted attempts of reconstitution. As will be described next, a design based on these considerations has allowed the successful reconstitution of the H^+ pump of clathrin-coated vesicles. Recent experiments (D. L. Schneider, X.–S. Xie, and E. Racker, unpublished observation) showed that the lysosomal H^+-ATPase has properties similar to those of the clathrin-coated vesicles with respect to solubilization, stimulation by phosphatidylserine and reconstitution.

C. Clathrin-Coated Vesicles

The role of these organelles in the internalization of toxins, viruses, hormones, receptors, and other proteins has been extensively reviewed (Goldstein *et al.*, 1979; Pearse and Bretscher, 1981; Pastan and Willingham, 1981; Ashwell and Harford, 1982; Helenius *et al.*,

1983). There was ample early evidence that the low-intravesicular pH in endocytic vesicles was involved in the internalization process, but direct evidence for an ATP-driven proton pump was only presented recently by Stone *et al.* (1983a) and Forgac *et al.* (1983). The pump bears some resemblance to the proton pump of mitochondria, but it is clearly distinct. It has an absolute requirement for Cl^- or Br^- as co-ions (Xie *et al.*, 1983). Phosphate, sulfate, or gluconate did not substitute for Cl^- or Br^-. Oligomycin, at concentrations that completely inhibited the mitochondrial proton pump, had only little effect on the pump of clathrin-coated vesicles, nor did several other mitochondrial inhibitors such as efrapeptin or azide affect proton translocation (Stone *et al.*, 1983a). DCCD at relatively high concentrations blocked proton translocation. NEM at 1 mM concentration had no effect on the proton pump of submitochondrial particles but completely inhibited the proton pump of clathrin-coated vesicles. When the Cl^- transport was blocked by duramycin or appropriate concentrations of detergents, H^+ translocation ceased but was restored by a valinomycin-induced efflux of K^+. An ATP-dependent uptake of $^{36}Cl^-$ was shown to be secondary to the influx of H^+, which generated an inside positive membrane potential. For Cl^- uptake, valinomycin-induced influx of K^+ substituted effectively for ATP (Stone *et al.*, 1984b).

Duramycin was discovered as an inhibitor of Cl^- transport during a systematic testing of about 150 antibiotics that had been stored away by Dr. Henry Lardy because of their weak or nonexisting effect on mitochondria. Among these antibiotics, which were generously supplied to us by Dr. Lardy together with a note saying "You will be sorry you asked," was a second component supplied by Eli Lilly called A20457, which was also found to be effective. In response to an inquiry about this drug, Dr. R. Hammill revealed that it was identical with duramycin. I shall return to a discussion of duramycin in the lecture dealing with glycolysis in tumor cells, but I want to emphasize here that duramycin cannot be used as a specific inhibitor of endocytosis, because it permeabilizes cells at low concentrations (Racker *et al.*, 1984). At somewhat higher concentrations it inhibits other processes in addition to Cl^- transport including the Na^+,K^+-ATPase (Stone *et al.*, 1984b; Nakamura and Racker, 1984).

The proton pump of clathrin-coated vesicles has recently been successfully reconstituted into liposomes in our laboratory using a

NEM-treated pellet together with a solubilized and purified preparation of the ATPase from coated vesicles (Xie *et al.,* 1984). Cholate dilution was the only reconstitution procedure successful thus far. These observations are reminiscent of the reconstitution of the mitochondrial proton pump with an insoluble preparation of F_0 and solubilized F_1 except that the ATPase of coated vesicles is much more hydrophobic than F_1. Since the chloride transporter was not co-reconstituted, detection of the successful reconstitution of the electrogenic ATP-driven proton pump of coated vesicles required generation of an inside negative membrane potential by an efflux of K^+ in the presence of valinomycin. This emphasizes that, for a successful reconstitution of an electrogenic transporter, an appropriate system for collapsing the membrane potential needs to be constructed.

III. ATP-Driven Ion Pumps of Microorganisms and Plants

A. The Proton Pumps of Microorganisms

I shall not deal here with the F_1F_0 proton pumps that were discussed in Lecture 5. However, I would like to reemphasize that this electrogenic pump functions in bacteria either as a proton excretor, e.g., under anaerobic conditions, or as an ATP generator under aerobic conditions. In contrast, the proton pump of the plasma membrane of yeast and *Neurospora* is of the E_1E_2 type, forming a phosphorylated protein intermediate, and its sole function is to excrete protons that are secondarily used to import nutrients by a proton symport mechanism. In this respect it fulfills some of the secondary functions of the Na^+, K^+ pump in mammalian cells.

In a comprehensive review, Goffeau and Slayman (1981) have covered this subject. I shall deal mainly with aspects relevant to reconstitutions. However I would like to raise a question of physiology mentioned earlier in the discussion of the stomach pump. Can the E_1E_2 pump alone account for the massive excretion of protons that had been observed over a hundred years ago by Lavoisier and later studied by Pasteur? Or is there another parallel pump, perhaps driven by a redox catalyst in the plasma membrane as was proposed by Conway (1951), 10 years before the beginning of the chemiosmotic storm? We still have no explanation for the almost ubiquitous presence of redox carriers in plasma membranes. Their possible participation in transport processes across the plasma membrane needs exploration.

There are quite a few novel observations that have emerged from the work on the resolution and reconstitution of the E_1E_2 proton pumps of yeasts and other fungi. The problem of cell breakage in the presence of a cell wall have been overcome by using either a wall-less mutant, in the case of *Neurospora* (Scarborough, 1978), or by digestion of the cell wall with snail gut enzyme (Bowman *et al.*, 1981). For the isolation of plasma membrane, a novel procedure using concanavalin A was devised (Scarborough, 1978). The cell surface was coated with the lectin to stabilize the plasma membrane and to facilitate its separation from other membranes. Concanavalin A was then dissociated with α-methylmannoside, thereby converting the membrane sheets into vesicles. Unfortunately, the vesicles isolated by this elegant method still contained considerable amounts of mitochondrial F_1F_0, complicating early reconstitution experiments (G. A. Scarborough and E. Racker, unpublished observation). In purified ATPase preparations from *Neurospora* digested by snail gut enzyme, mitochondrial ATPase was barely detectable (Bowman *et al.*, 1981). Alternatively, mechanical breakage followed by precipitation of mitochondrial membranes at an acid pH of 4–5 (Schneider *et al.*, 1978) and by sucrose gradient fractionation have yielded membranous ATPase preparations from yeast, free of mitochondrial ATPase (Dufour and Goffeau, 1978). The specific activity of these membranes varied in different preparations from 1.0 to 8.0 μmol $\times$ min^{-1} $\times$ mg protein^{-1}. The enzyme was found to be specific for ATP (or dATP). It was inhibited by vanadate and DCCD but not by azide or oligomycin. Stripping these membranes of "impurities" at low detergent concentration yielded a preparation of specific activity of 20 or higher. A large-scale preparation from *Neurospora* had a specific activity of 35, which is somewhat higher than that of most E_1E_2-ATPases from mammalian tissues (Smith and Scarborough, 1984). Although the best preparations revealed a single band at 100,000 daltons in SDS–PAGE, the possibility of the presence of a proteolipid is difficult to rule out as discussed earlier.

Of particular interest are observations on enzyme activation by phospholipids (Dufour and Goffeau, 1980; Bowman *et al.*, 1981). The purified enzyme tended to aggregate, exhibiting little ATPase activity, but could be restored by addition of phospholipids. Lysolecithin was particularly effective, whereas changes in the head group had only a minor influence. Even dilauryllecithin was quite effective. Similar data on the effect of various phospholipids on proton pumping would be of considerable interest. As was shown in the case of the Ca^{2+} pump (Knowles *et al.*, 1975; Navarro *et al.*, 1984), the effects of phospholipids on ATPase activity may profoundly differ from those on transport activities.

The mechanism of action of these pumps is quite similar to that of other E_1E_2-ATPases, including multiple binding sites for Mg^{2+} (Ahlers *et al.*, 1978) and a slow rate of $^{32}P_i$–ATP exchange (Malpartida and Serrano, 1981b). These pumps are, like other ATP-driven pumps that have been carefully analyzed, electrogenic.

B. The ATP-Driven Ca²⁺ Pump in Streptococcus faecalis

Membrane fragments of *S. faecalis* prepared by disruption of the cells in a French press are a mixture of right side out and inside out vesicles. The latter catalyze an ATP-driven uptake of Ca^{2+} (Kobayashi *et al.*, 1978). The process takes place via a phosphorylated intermediate and is exquisitely sensitive to vanadate (Bürkler and Solioz, 1982). It is therefore surprising that, unlike in other typical E_1E_2 enzymes, the process was found to be insensitive to DCCD. However, this property should be reexamined in view of the observation that the activity of Ca^{2+}-ATPase of SR was only DCCD-sensitive after removal of Ca^{2+} bound to a hydrophobic domain (Pick and Racker, 1979b). The bacterial enzyme was reconstituted into crude soybean phospholipids in the presence of Triton X-100 followed by the removal of the detergent on a column of amberlite XAD-2. The properties of the transport in the reconstituted proteoliposomes were the same as in the native vesicles. The Ca^{2+} uptake following reconstitution and the vanadate-sensitive ATPase activity were used as assays during the purification of the solubilized enzyme via a glycerol gradient. Surprisingly, the vanadate-sensitive ATPase was not inhibited by EGTA, suggesting that more than one ATP-driven transporter is present in this fraction. Indeed transport of both Na^+ and K^+ appear to be ATP-dependent in these organisms (Heefner *et al.*, 1980; Bakker and Harold, 1980). These studies on the Ca^{2+} pump of *S. faecalis* illustrate once again that the best assay for a transporter protein is reconstitution. The multiplicity of ATPases in membranes makes the assay of ATP hydrolysis ambiguous even when an inhibitor, such as vanadate, is available. It should be remembered that some nonspecific phosphatases are sensitive to vanadate (O'Neal *et al.*, 1979).

C. The ATP-Driven Pumps for K⁺ and Other Ions

As mentioned before, K^+ transport in *S. faecalis* is ATP-dependent and in *E. coli,* one of the K^+ transporters has been shown to be ATP-dependent (Rhoads and Epstein, 1977). The ATPase activity of this pump is sensitive to vanadate (O'Neal *et al.*, 1979). There are numerous indications that other bacterial transport systems, including some amino acids and P_i, are driven by ATP with or without the participation of the proton motive force. Thus far, none of these interesting transport systems appears to have been reconstituted.

D. The ATP-Driven H⁺ Pumps of Plants

Electrogenic proton pumps in plants play a role in the transport of inorganic ions, sugars, and amino acids and in the regulation of growth and development (Poole, 1978; Spanswick, 1981). Two distinct pumps have been described in membrane preparations from various plants, particularly red beets (Bennett *et al.*, 1984) and oat roots (Churchill *et al.*, 1983). One located in the plasma membrane appears to be a vanadate-sensitive pump of the E_1E_2 type; the second present in tonoplasts, is a vanadate-insensitive pump. The latter was solubilized from corn roots by deoxycholate in the presence of 40% glycerol and reconstituted into purified phospholipids. A deoxycholate-solubilized and partially purified vanadate-sensitive ATPase from plasma membranes of red beets was reconstituted with soybean phospholipids (O'Neill and Spanswick, 1984). A highly purified vanadate-sensitive ATPase, solubilized from oat roots with lysolecithin, was reconstituted with soybean phospholipids by freeze–thaw sonication (Vara and Serrano, 1982; Serrano, 1984). In general these vanadate-sensitive pumps resemble the H^+ pump in plasma membranes of yeast, whereas the tonoplast pump resembles that of clathrin-coated vesicles.

I have on purpose included in this lecture pumps of the E_1E_2 type in microorganisms and plants that could have been discussed in Lecture 6, together with the plasma membrane pumps of mammalian cells. But since investigators in the field of microorganisms and plants rarely read papers on mammalian cells, I thought it best to keep the discussion separate.

Lecture 8

Proton Motive Force Generators, Electron Transport Chains, and Bacteriorhodopsin

> You can't depend on your eyes
> when your imagination is out
> of focus.
>
> **Mark Twain**

I. Reconstitution of the Mitochondrial Electron Transport Chain

The first resolution of the mitochondrial electron transport chain into four complexes was achieved in the laboratory of David Green by Hatefi *et al.*(1962). These complexes have now been reconstituted into artificial phospholipid bilayers. The first example was cytochrome oxidase, which was reconstituted by cholate dialysis into unilamellar liposomes together with the proton pump (Racker and Kandrach, 1971) as well as alone (Hinkle *et al.*, 1972). It was incorporated into planar, thick membranes made with soybean phospholipids by Jasaitis *et al.* (1972). Subsequently, many other electron transporters from mitochondria, chloroplasts, and bacteria were reconstituted using a variety of procedures. Many of these experiments have been discussed previously (Racker, 1976). In a chapter devoted to reconstitutions in a book on "Membrane Bioenergetics" (C.-P. Lee *et al.*, 1979) Nelson, Hauska,Trumpower, Rydström, and others deal with reconstitution developments in this area in the subsequent three years. Specialized reviews on isolated complexes have appeared since (Trumpower, 1981; Azzi, 1980; Ragan *et al.*, 1981; Wikström *et al.*, 1981a,b; Capaldi *et al.*, 1983; Hauska *et al.*, 1983). I shall concentrate here on some of the more recent developments that have emerged from reconstitution experiments.

A. Cytochrome Oxidase

1. Subunit Structure

This enzyme has been isolated in a reconstitutively active form from mammalian tissues, yeast, and bacteria, yet its subunit composition is still controversial. In a recent review (Capaldi *et al.*, 1983) the point was made that, with increased resolutions achieved by ever-improving SDS–PAGE, the numbers of subunits have increased with time, both in the case of the yeast and mammalian cytochrome oxidase. It was properly pointed out that subunit composition must be defined. If defined in terms of the polypeptides that copurify with heme and copper in stoichiometric amounts, then indeed the number of subunits may be as high as 12–13. Some of the difficulties in accepting such a definition were discussed by Azzi (1980). I prefer in principle an alternative, functional definition based on the minimal number of subunits required for function, i.e., electron transport with respiratory control, with and without proton pumping. According to this definition, the answer should be that only two subunits are essential, as demonstrated in the case of cytochrome oxidase from *Paracoccus denitrificans* (Ludwig and Schatz, 1980; Ludwig, 1980). The apparent molecular weight of the large subunit (subunit I) is 45,000; of the smaller (subunit II) 28,000. The latter cross-reacts immunologically with subunit II of the yeast enzyme. Some other bacterial cytochrome oxidases appear to be similar, though somewhat more complex (Sone, 1981; Sone and Yanagita, 1982).

Mammalian cytochrome oxidase from bovine heart mitochondria is clearly more complex than the bacterial one. In our own experience, extensive treatment of this enzyme with chymotrypsin yielded a preparation showing only six major bands in SDS–PAGE, yet exhibiting respiratory control and catalyzing proton pumping (Carroll and Racker, 1977; Saltzgaber-Müller *et al.*, 1980). Therefore, it seems likely that the other subunits are either fellow SDS travelers or only peripherally concerned with cytochrome oxidase function (e.g., regulation or assembly). Amino acid sequencing and topology of subunits, as well as the electron microscope work of Henderson and his associated (1977) on the structure of cytochrome oxidase, have been discussed in other reviews (Azzi, 1980; Capaldi *et al.*, 1983). Perhaps similar image reconstruction analysis of the simpler *Paracoccus denitrificans* enzyme will yield information on the role of the two subunits.

The mitochondrial cytochrome oxidase of yeast is also more complex than the bacterial enzyme (Schatz and Mason, 1974). Present evidence points to seven subunits, with the three large subunits being synthesized by the mitochondria, the four smaller subunits by cytoplasmic ribosomes, and imported into mitochondria.

The definition of functionality also has complexities, since respiratory control and even proton pumping may be insufficient criteria for *in vivo* function. Indeed the same objections can be raised to a definition based on genetic analysis, which may miss subunits that are involved in subtle control mechanisms that operate under unphysiological conditions, e.g., at high temperature or other unfavorable conditions such as the presence of environmental toxic substances. In any case, a definition based on guilt by association, as proposed by Capaldi *et al.* (1983), may have its usefulness in terms of calling for additional work.

Lesson 26: Limits of definition of subunit compositions

A definition is a tool for the purpose of facilitating the clarity of communication. If definitions lead to answers that are as discrepant as in the case of cytochrome oxidase, varying with the methods of purification and analysis, they become useless and lead to confusion rather than to clarification. Since we obviously cannot communicate without defining what we mean when we speak of cytochrome oxidase, we must specify in each case the limits of our analysis. Then we can state that cytochrome oxidase$_{SDS}$ has twelve subunits or more, cytochrome oxidase$_{gen}$ at least three subunits, and cytochrome oxidase$_{func}$ may have as few as two. Since all three definitions call for more experiments, we should be glad to accept them all.

2. Reconstitutions of Electron Transport, Respiratory Control, and Proton Pumping with Cytochrome Oxidase

Perhaps the most striking aspect of reconstitutions of cytochrome oxidase is the previously mentioned fact that it preferentially results in the formation of right side out vesicles. Once again we can explain this on the basis of its subunit topology (Henderson *et al.*, 1977). A relatively large proportion of cytochrome oxidase on the C-side of the membrane (facing the cytosol) is in the water phase. If this is indeed the explanation for the preferential insertion of the more hydrophobic M- (matrix) side of the enzyme into the phospholipid bilayer, it is curious that in the presence of oxidized cytochrome c (but not of reduced cytochrome c), a large fraction of the enzyme (ca. 40%) is reconstituted in the inside out configuration (Carroll and Racker, 1977). Is it likely that the hydrophilic cytochrome c itself serves as a pilot protein for the insertion of the enzyme, or does the interaction with cytochrome c render the C-side more hydrophobic or the M-side more hydrophilic by a conformational change? Image-reconstruction experiments with the cytochrome oxidase-cytochrome c complex may shed light on this problem. The phospholipid composition, the phospholipid-to-protein ratio, and the method of reconstitution appear to influence the size, the respiratory control, and orientation of cytochrome oxidase vesicles (Madden *et al.*, 1984; Madden and Cullis, 1985). The smaller the vesicles, the higher the respiratory control ratio. A large excess of phospholipids, which favors the presence of only one dimer per vesicle, was optimal for respiratory control. Of special interest are the observations that right side out and inside out cytochrome oxidase vesicles can be separated either by DEAE chromatography or by a cytochrome c affinity column.

It may be appropriate to review here briefly the history of the pump activity of cytochrome oxidase, since it documents the value of reconstitution experiments. Acidification of the media on addition of reduced cytochrome c to reconstituted cytochrome oxidase from bovine heart mitochondria was first observed by Hinkle (1973). Since the effect was very small, involving less than one proton per electron, it was considered to be an artifact; a view held until recently by Mitchell. Wikström and Sari (1977), Wikström *et al.* (1981a,b), Sigel and Carafoli (1980), and Solioz *et al.* (1982) have accumulated forceful evidence in favor of the operation of a proton pump catalyzed by cytochrome oxidase. A careful examination of this problem (Coin and Hinkle, 1979), with reconstituted cytochrome oxidase vesicles that had a respiratory control ratio of 10, confirmed the conclusion drawn by Wikström. Under conditions of less than three turnovers of the enzyme, $H^+/2e^-$ ratios in excess of 2 were observed. With very small oxygen pulses, allowing for less than one turnover of cytochrome oxidase, the $H^+/2e^-$ ratios approached 4.0.

Perhaps a better understanding of the modes of interaction between cytochrome c and cytochrome oxidase will eventually help to solve these puzzling observations. Based on careful kinetic studies with purified cytochrome oxidase of the rates of reduction of a and a_3, it was suggested many years ago that the unexpected slow rate of electron transfer from a to a_3 with the isolated enzyme is due to the formation of an inhibitory complex between oxidized cytochrome c and reduced cytochrome a (Gibson et al., 1965). Continuous electron flux was explained by a slow dissociation of oxidized cytochrome c and its replacement by reduced cytochrome c. This is probably not what happens in a functioning electron transport chain which delivers electrons to bound cytochrome c. In line with the formulation by Gibson et al. (1965) are the observations made on multiple-site interactions between cytochrome c and the oxidase (Ferguson-Miller et al., 1978). These investigators propose that cytochrome oxidase that binds one molecule of cytochrome c in position 1 is capable of binding a second molecule of cytochrome c in position 2 in a catalytically active manner with a K_D 50-fold higher than for position 1. Two binding sites in cytochrome oxidase for cytochrome c have been documented, one at subunit II (Azzi, 1980) and one at subunit III (Birchmeyer et al., 1976). It should be pointed out that the conditions for the measurement of proton pumping with reconstituted cytochrome oxidase do not include a physiological reducing system for cytochrome c while it is bound to cytochrome oxidase. If the oxidoreduction of cytochrome c in position 1 or 2 is essential for proton translocation, an appropriately reconstituted electron delivery system to bound cytochrome c might not only induce steady-state proton pumping, but may eliminate also the well-known, unphysiological requirement for high concentrations of cytochrome c. A cyclic, conformational change of cytochrome oxidase, associated with an induced increased hydrophobicity of its subunits located on the C-side of the membrane when cytochrome c becomes oxidized in position 1, may represent a contractile or conformational model for an operative proton pump analogous to that of bacteriorhodopsin (Fig. 8-1). Studies of conformational changes in cytochrome oxidase induced by oxidoreduction of cytochrome c of the type described for the interaction of cytochrome c with reconstituted photosynthetic reactions centers (Pachence et al., 1983) may yield some useful results. Recent observations (Wikström and Penttilä, 1983) with rat liver mitochondria revealed that proton translocation catalyzed by cytochrome oxidase was greatly enhanced at high salt concentrations, which presumably favor the release of cytochrome c from the high-affinity site in cytochrome oxidase. Since TMPD also increased proton pumping, these experiments are consistent with the formulations for the proton pump activity requiring in situ cytochrome c reduction as outlined before.

Subunit III of cytochrome oxidase has been implicated in the proton pump mechanism (Wikström 1981; Wikström et al., 1981a,b). This subunit could be removed without significantly altering electron transfer activity, and vesicles reconstituted with this preparation exhibited good respiratory control. However, these proteoliposomes did not pump protons. Earlier evidence for the involvement of subunit III has emerged from studies on the effect of DCCD (Azzi et al., 1979, Casey et al., 1980). This compound was shown to interact specifically with subunit III and to inhibit proton pumping in reconstituted cytochrome oxidase vesicles. The observation that a prolonged exposure (3 h) of the enzyme to low concentrations of DCCD inhibits proton pumping appears to eliminate the

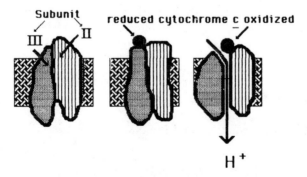

Fig. 8-1 A conformationl model of the cytochrome oxidase proton pump.

participation of an uncoupler that is present in some DCCD preparations (cf. Coin and Hinkle, 1979).

A bacterial cytochrome oxidase lacking subunit III was shown to be capable of proton pumping (Solioz *et al.*, 1982). Another bacterial *b*-type cytochrome oxidase from *Rhodopseudomonas capsulata* was reconstituted into phospholipid vesicles by the cholate dialysis procedure (Hüdig and Drews, 1984). Whereas crude preparations showed H^+ pumping activity (0.9 H^+/e^-) that was sensitive to DCCD, a highly purified two-subunit preparation did not. Of particular interest is the observation that DCCD did not inhibit respiration, suggesting that the system was "loosely coupled" or was subject to molecular "slipping" to use the more fashionable term (Wikström *et al.*, 1981a,b). It will be interesting to explore by reconstitution experiments what was lost on the way to the purified two-subunit enzyme.

Two significant publications have recently appeared, bearing on the role of subunit III in mammalian cytochrome *c* oxidase. The observation that chymotrypsin-treated cytochrome oxidase from bovine heart is still capable of pumping H^+ (Saltzgaber *et al.*, 1980) was confirmed (Puettner *et al.*, 1985). These authors, moreover, showed by labeling of subunit III with [14C]DCCD that digestion of this subunit has taken place. The reconstituted enzyme operated as a H^+ pump though with some diminished efficiency.

Cytochrome oxidase extracted from rat liver with laurylmaltoside and purified by cytochrome *c* affinity chromatography lacked subunit III (Thompson, 1985). The reconstituted enzyme exhibited respiratory control and pumped protons.

These two important contributions need further explorations and the possibility of a nicked subunit III must be rigorously excluded. Further studies with protease inhibitors present during purification of cytochrome oxidase may be revealing.

B. The Cytochrome b-c₁ Segment (QH₂-Cytochrome c Reductase, Complex III)

1. Isolation, Composition, and Function

Purification of QH_2-cytochrome *c* reductase has been achieved by a variety of methods with different detergents. Useful summary tables of these procedures and of the composition of the various preparations have been recorded in an excellent review (Hauska *et al.*, 1983). Usually 7–8 major polypeptides are seen in SDS–PAGE. Whereas the two core proteins, cytochrome *b*, cytochrome *c*, and the Rieske iron–sulfur protein are well-established members of this complex, limited proteolysis contributes to some of the polypeptides smaller than 15,000 daltons, seen in preparations of the yeast complex (Sidhu and Beattie, 1982). Studies on the biogenesis of the complex in *N. crassa* suggest that 8 subunits, including some smaller than 15,000 daltons, are synthesized as precursors and assembled into the complex after being transported into the mitochondria (Teintze *et al.*, 1982).

Numerous reviews on the function of the individual components and their role in the Q cycle have been published (Hauska *et al.*, 1983; van Jagow and Sebald, 1980; Trumpower, 1981; King, 1981; Slater, 1983). Based on diffraction patterns of QH_2-cytochrome *c* reductase from *N. crassa*, a model was constructed in which a large portion of the complex protrudes from the membrane into the mitochondrial matrix (Weiss *et al.*, 1979).

2. Resolution and Reconstitution

A first step in the physical resolution and reconstitution of complex III was achieved by Nishibayashi *et al.* (1972), revealing the functional requirement of an "oxidation factor" which was later identified as the native form of the Rieske iron–sulfur protein (Trumpower and Edwards, 1979). Precautions required for the isolation of a depleted complex and its reassembly with the reconstitutively active iron–sulfur protein have been extensively described (Trumpower, 1981). A stepwise resolution of the complex from *Neurospora* was also achieved (Weiss *et al.*, 1979). Of particular interest are reconstitutions of complex III with other members of the electron transport chain. Starting with the classical experiments of Hatefi *et al.* (1962), reconstitutions with more highly purified complexes from mammalian, plant, and microbial membranes have been accomplished (Hauska *et al.*, 1983). Coenzyme Q, discovered as a cofactor of respiration in the laboratory of David Green, made subsequent assays during reconstitution of complex III possible.

Reconstitution of oxidative phosphorylation by co-reconstitution of complex III with mitochondrial F_1F_0 was achieved by cholate dilution (Racker *et al.*, 1975). The first functional reconstitution of complex III into liposomes catalyzing respiration with respiratory control was achieved by cholate dialysis (Leung and Hinkle, 1975). The observed stoichiometry in the reconstituted vesicles of about two for $H^+/2e^-$ was similar to that observed with mitochondria. The same complex isolated from yeast mitochondria gave a similar uncoupler-sensitive ratio for proton translocation (Beattie and Villalobo, 1982). In these vesicles DCCD blocked the proton ejection process when used at concentrations that did not affect the proton permeability of the liposomes. Nalecz et al. (1983), however, concluded from experiments with reconstituted bovine QH_2-cytochrome c reductase that both electron transport and H^+ translocation were inhibited in the reconstituted particles following DCCD treatment and they correlated the inhibition with the cross-linking of two subunits (V and VII) of the complex. They also concluded that the observed binding of DCCD to cytochrome b was not kinetically correlated with the observed inhibition. In more recent studies Clejan and Beattie (1983, and personal communication) have shown that under appropriate conditions (at 12°C) DCCD blocked proton fluxes with minimal changes in electron transport in reconstituted complex III of yeast mitochondria.

C. NADH Dehydrogenase (Complex I)

Purified complex I contains FMN, nonheme iron, acid-labile sulfide, and a bewildering number of proteins and of iron–sulfur centers (Ragan et al., 1981; Ohnishi, 1979). Although the flavoprotein and the iron protein have been separated as discussed previously (Racker, 1976), further resolution experiments have to be performed to establish the need and function of over 20 polypeptide chains and at least 7 iron–sulfur centers found in complex I.

There is little known about the mechanism of the proton pump associated with complex I (Ragan et al., 1981). Although Mitchell has proposed (1979) a Q cycle also for this respiratory segment, the absence of Q from some preparations of NADH dehydrogenase casts doubts on this formulation (cf. Ragan et al., 1981). However, since these particular preparations of complex I have not as yet been shown to be reconstitutively active in a system capable of supporting ATP formation in the presence of F_1F_0, the possibility of Q participation has not as yet been eliminated. On the other hand, a contractile or conformational proton pump mechanism such as suggested for cytochrome oxidase could be visualized to operate in this segment also. A direct approach to this problem would require experimental evidence for pronounced conformational changes of the enzyme during interaction with its reduced and oxidized substrate. Such experiments might be feasible with highly purified preparations that can be incorporated into artificial liposomes (Ragan and Hinkle, 1973).

D. The Nucleotide Transhydrogenase

Nicotinamide nucleotide transhydrogenase catalyzes the reversible hydride transfer between $NADH_2$ and NADP. The enzyme is present in bacteria and in the inner mitochondrial membrane. I discuss it in this lecture because it is involved in oxidoreduction and translocation of protons. An excellent historical review of the reactions catalyzed by this enzyme and the controversies over its mode of action has been published (Rydström, 1977).

The first experiments on reconstitution from without showed that energy-driven reduction of NADP was dependent on the addition of F_1 to resolved submitochondrial particles (Fessenden-Raden, 1969). An interesting stimulation by P_i of the reaction driven by oxidative energy does not seem to have been further elucidated. In bacteria there are several other solubilized factors that are needed for the transhydrogenase (Rydström, 1977). A reconstitution of the ATP-driven transhydrogenase into liposomes with partially purified components prepared from complex I of bovine heart mitochondria together with F_1F_0 was reported by Ragan and Widger (1975). A partially purified transhydrogenase of bovine mitochondria was reconstituted into phosphatidylcholine liposomes without F_1F_0 (Rydström et al., 1975). In these particles the reduction of NAD by $NADPH_2$ generated a membrane potential, as indicated by the uncoupler-sensitive uptake of tetraphenyl boron, a lipophilic anion. Moreover, NAD reduction was greatly stimulated by proton ionophores.

Experiments from the laboratories of Rydström and Fisher with highly purified and reconstituted transhydrogenase (Anderson and Fisher 1978, 1981; Rydström et al., 1981; Enander and Rydström, 1982) have shown that the enzyme is composed of a single hydrophobic polypeptide chain with a molecular weight of about 110,000. Reconstituted into liposomes or into planar membranes, the enzyme generated an uncoupler-sensitive proton motive force, translocating one proton for each NAD reduced by $NADPH_2$. Once again, inhibition by DCCD (Phelps and Hatefi, 1981; Pennington and Fisher, 1981) has been interpreted in terms of the operation of a proton pump (Skulachev, 1974). When NAD was reduced with either NADH or $NADPH_2$ as hydride ion donor, protons were translocated from the M-side to the C-side; when NADP was reduced, the H^+ flux was from the C-side to the M-side. It was shown (Pennington and Fisher, 1981) that 1 mole of DCCD per mole of enzyme inhibited proton translocation, whereas 2 moles of DCCD per mole of enzyme resulted in the cross-linking of the dimer and inhibition of hydride transfer between the nucleotides. It was concluded that the inhibition of proton flux by DCCD takes place by modification of a site other than the active site involved in hydride transfer.

Unlike other electron catalysts of the mitochondrial membrane, homogeneous preparations of transhydrogenase can be reconstituted with either dioleoylphosphatidylcholine or dioleoylphosphatidylethanolamine (Rydström, 1979). Acidic phospholipids such as cardiolipin were strongly inhibitory even when present at concentrations less than 10% of total phospholipids. In view of the presence of 20% cardiolipin (of total phospholipids) in the inner mitochondrial membrane, this inhibitory phenomenon needs further exploration. It is conceivable that either other lipids or protein components modify this inhibition or that the method of reconstitution does not mimic the natural distribution and orientation of the phospholipids. It would be of interest to learn whether cardiolipin inhibits equally when different methods of reconstitution are used.

Proton translocation in reconstituted vesicles was measured by the quenching of the fluorescence of 9-amino-6-chloro-2-methoxy-acridine. Intravesicular acidification was achieved by collapsing the membrane potential generated during the reaction with sodium nitrate. A co-reconstitution with F_1F_0 at a 4:1 ratio of phosphatidylethanolamine /phosphatidylcholine allowed the demonstration of an ATP-dependent stimulation of the reduction of NADP by $NADH_2$.

E. Oxidoreductions in Bacteria

The panorama of different electron transport catalysts in bacteria together with the opportunities of nutritional and genetic manipulation on one hand, and the scarcity of ~ur knowledge of the reason for these divergencies of catalysts on the other hand make this a challenging field for reconstitution experiments. The few reconstitution experiments with oxidative enzymes of *E. coli,* thermophilic bacteria, and *Paracoccus denitrificans* mentioned earlier have already demonstrated the value of this approach. It is therefore remarkable how little has been done on the reconstitution of bacterial oxidative enzymes. Particularly instructive is the example of *P. denitrificans,* which has been looked upon as a promising candidate for the role of an ancestor of mitochondria in eukaryotes (John and Whatley, 1977). The similarities between the oxidative catalysts of the plasma membrane of these bacteria with the inner membrane of mitochondria are indeed remarkable. The two systems exhibit the same sensitivity to inhibitors (not seen with many other microorganisms): the presence of Q_{10} and phosphatidylcholine as major lipid constituents and oxidoreduction catalysts with similar kinetic properties. In spite of these apparent similarities, the approach of resolution and reconstitution of cytochrome oxidase from these bacteria has yielded important and unexpected information. In contrast to cytochrome oxidase from mitochondria, the enzymes from *P. denitrificans,* from thermophilic PS3, and several other bacteria (Ludwig, 1980; Sone *et al.,* 1983) have a very simple subunit composition. The PS3 cytochrome oxidase with 3 subunits (subunit II containing cytochrome *c*) was shown to pump protons with a high H^+/O ratio (Sone and Yanagita, 1982). In the case of *P. denitrificans,* the enzyme contains only two polypeptides of M_r 45,000 and 28,000 (Ludwig and Schatz, 1980), corresponding to subunit I and II of mitochondria. Yet it appears from reconstitution experiments that the bacterial enzyme has all the functional properties of the reconstituted mitochondrial oxidase, including a high respiratory control and the ability to pump protons (Solioz *et al.,* 1982). As in the case of mitochondrial oxidase, the pump appeared to function for a limited number of turnovers (2–12), and extrapolation to zero turnover yielded a ratio of only 0.6 protons per electron. I have discussed the problems associated with this phenomenon in the section dealing with the eukaryotic enzyme. They need further exploration, and the finding of proton pumping in a reconstituted cytochrome oxidase that does not contain the DCCD binding subunit III of the eukaryotic enzyme requires a reevaluation of the role of the subunit in the pumping process.* The simplicity of the bacterial oxidase also raises the question

* See page 130 for more recent developments.

of the roles of the smaller subunits in the eukaryotic enzyme and points to some secondary regulatory functions or participation in the proper assembly during biogenesis.

Another example of a reconstitution of bacterial electron transport is the reduction of fumarate by formate in *Vibrio succinogenes* (Unden and Kröger, 1982; Unden *et al.*, 1983). The complex catalyzing this reaction consists of formate dehydrogenase (3 different subunits) and fumarate reductase (3 different subunits) and uses vitamin K-1 as a redox component. The electron transport system was incorporated into PC liposomes by an octylglucoside dialysis procedure developed by Helenius *et al.* (1977), followed by freezing and thawing. Of particular interest is the mode of incorporation and the presence of peripheral hydrophilic projections seen in electron micrographs. The future possibilities are elucidations of the role of individual subunits, their location in the lipid bilayer, and the mechanism by which this electron transport chain is associated with the formation of ATP, which takes place in the native membrane.

An excellent example for the advantages of using bacteria for the characterication and reconstitution of electron transfer enzymes is fumarate reductase of *E. coli* (Lemire *et al.*, 1983; Weiner *et al.*, 1984). Based upon a combination of genetic, physical, enzymatic, and electron microscopic studies, a model was suggested which is likely to be relevant for other iron–flavo–enzyme complexes, e.g., in mitochondria. Great enrichment of the enzyme was achieved by growing an *E. coli* strain that carried a multicopy plasmid for the reductase operon under anaerobic conditions in the presence of fumarate and glycerol. The enzyme is a membrane-bound tetramer consisting of four nonidentical subunits: a 69,000-dalton subunit containing FAD, a 27,000-dalton subunit containing a nonheme iron–sulfur center, and two small very hydrophobic subunits that are embedded in the membrane. On treatment with 6 M urea, the large subunits were removed. By adding back the large subunit in the presence of DTT, both functional and morphological (knob-like) structures) reconstitution was achieved. The reconstituted particles looked remarkably similar to submitochondrial particles with a head group (69K subunit) and a stalk (27K subunit). Unpublished experiments from this research group suggest that the components of this system can be resolved and reconstituted and may be easier to manipulate than other bacterial proton pumps.

II. Reconstitution of Photosynthetic Electron Transport Pathways

As pointed out in illuminating reviews by Nicholls (1981) and Hauska *et al.* (1983), chloroplasts differ from mitochondria in several significant aspects that are relevant for reconstitutions: (1) the structure of chloroplasts is more complex, with the energy transducing system of CF_1F_0 in the opposite orientation from that in mitochondria; (2) chloroplasts contain large quantities of light-harvesting pigments that participate by excitation transfer; (3) they contain a mechanism of water cleavage that is unique. The electrons released by this cleavage channel into an electron transport system that is basically similar to that in mitochondria.

Since earlier work was reviewed by Nicholls (1981), I shall only touch upon some highlights and more recent work. Of particular interest is the reconstitution of the light-harvesting complex and the demonstration of protein aggregation on addition of Mg^{2+} (Mullet *et al.*, 1981). In innovative reconstitution experiments Skulachev and his collaborators (Barsky *et al.*, 1976) have shown that proteoliposomes containing plant or bacteriochlorophyll complexes can be fused in the presence of Ca^{2+} with planar membranes. In an imaginative experiment with vesicles containing both bacterial reaction centers as well as bacteriorhodopsin fused to planar membranes, electrical measurements showed generation of electric fields that were dependent on the spectral composition of the light. Red light, exciting bacteriochlorophyll, induced negative, and green light, exciting bacteriorhodopsin, induced positive charging of the proteoliposome interior. Association of isolated *R. rubrum* chromatophores with planar membranes yielded, in the presence of napthoquinone, phenazine methosulfate, and ascorbate, a membrane potential as high as 215 mV. In this reconstituted system, addition of ATP or pyrophosphate also generated a membrane potential. Similar experiments have been conducted with cytochrome oxidase (Drachev *et al.*, 1976a) and with F_1F_0 of mitochondria (Drachev *et al.*, 1976b).

Important progress has been made (Nelson and Hauska, 1979) in the purification and reconstitution of components of the inner membrane of plant chloroplasts that contain three major functional units, as shown schematically and simplified in Fig. 8-2. As pointed out previously (Racker, 1976), an analogy can be drawn with the mito-

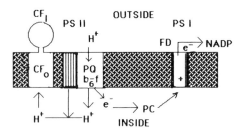

Fig. 8-2 Electron and proton fluxes in chloroplasts.

chondrial electron transport chain: PSI takes the place of cyto-chrome oxidase, cytochrome b_6-f corresponds to cytochrome b-c_1, plastocyanine to cytochrome c (cytochrome c 552 functions like plastocyanine in *Euglena* chloroplasts), and plastoquinone to coen-zyme Q. The pathway of electrons flows from PS II to cytochrome b_6-f complex to plastocyanine to PSI. The protons come from oxida-tion of water and are shuttled via quinones.

This scheme has gained confirmation from studies with reconsti-tuted systems. Purified photosystem I reaction centers from spinach chloroplasts (Bengis and Nelson, 1975; Orlich and Hauska, 1980; Hauska *et al.*, 1980) catalyze a high rate of NADP reduction (1800 μmol $\times$ mg chlorophyll^{-1} $\times$ hr^{-1}) in the presence of plastocyanine, ferredoxin, and ferredoxin–NADP oxidoreductase. These reaction centers were reconstituted in the presence of carefully controlled concentrations of Triton X-100 by direct sonication (15–30 min) into soybean phospholipids to yield unilamellar liposomes that were sta-ble for several days. The vesicles catalyzed proton extrusion in the light in the presence of reduced PMS, but also catalyzed light-depen-dent proton uptake as measured by quenching of 9-aminoacridine fluorescence. Thus the reconstitution was scrambled, yielding both right side out and inside out vesicles. When the reaction center complex was reconstituted with CF_1F_0 from spinach, the proteoli-posomes catalyzed light-driven ATP generation, sensitive to uncou-pler and to DCCD at the rate up to 50 μmol $\times$ mg chlorophyll^{-1} $\times$ hr^{-1}. Several methods of incorporation of the CF_1F_0 complex into the reaction center liposomes were explored. Freeze–thaw with or without brief sonication was the least effective. Simple mixing and immediate dilution into the assay mixture was the most effective. Since Triton X-100 was present in the reaction center liposomes, the basic principle of reconstitution is presumably detergent dilution.

Optimal ATP synthesis was observed at pH 8.5 and was linear for 1 hr. It was realized, however, that the observed rates of the reconstituted system represent only a small fraction of the rate catalyzed by intact chloroplasts.

The photosynthetic oxygen evolution system has been a matter of controversy for many years. Recently, important progress has been made in the partial resolution and reconstitution of photosystem II, particularly in the laboratories of Murata and Yocum. Release of several subunits has been achieved by treatment of PSII particles with 1 or 2 M NaCl (Miyao and Murata, 1983; Ghanotakis et al., 1984). Of particular interest is the observation of the latter that addition of Ca^{2+} to salt-treated membranes from which the 17- and 23-kDa polypeptides had been removed, restored oxygen evolution. Two Mn atoms were released together with the 33-kDa protein by addition of Tris. This protein appears to play a key role in oxygen evolution (Murata et al., 1984), although it does not contain bound Mn (Ghanotakis et al., 1984).

Interactions between oxidoreduction complexes from different sources in unphysiological, reconstituted (or constructed) systems have often been fruitful, as pointed out in the first lecture. A more recent example is the reduction of spinach cytochromes b_6 and f by photochemical reaction centers from *Rhodopseudomonas sphaeroides* (Prince et al., 1982). The availability of the chlorophyll-free, plastoquinol–plastocyanine oxidoreductase complex from spinach allowed the performance of sophisticated optical measurements in a rapidly responding dual-wavelength spectrophotometer. This system provides an interesting model for an ''oxidant-induced reduction'' analogous to that observed in mitochondria: oxidation of cytochrome f is required for the reduction of cytochrome b_6 in a coupled reaction where both electrons must leave the two electron carriers.

Another interesting construction was made with mitochondrial cytochrome b-c_1 complex and the reaction center isolated from *Rhodopseudomonas sphaeroides* (Rich and Heathcote, 1983). The proteins were incorporated into soybean phospholipids by cholate dialysis against unbuffered sucrose-KCl preceeded by a 5-hr dialysis against 0.1 M P_i, thus providing a high internal buffering capacity. On addition of cytochrome c and CoQ_2, light activation resulted in proton excretion into the medium in the presence of valinomycin. Cytochrome c on the outside was reduced by the b-c_1 complex and oxidized by the reaction center on the outside of the vesicles. CoQ_2 was reduced by the reaction center on the inside. The quinol was reoxidized by the b-c_1 complex on the outside. The $H^+/2e^-$ ratio was close to 4.

Perhaps I have bored you by the mass of information that has been gathered on electron transport in various laboratories during the past few years. Perhaps you find it difficult to see the forest for all the trees. Yet electron transport is an important part of the ATP-generating machinery and has replaced in nature the more primitive and more direct way of inducing a proton gradient, e.g., by illumination of bacteriorhodopsin, which I shall discuss next. What comes across is the astonishing number of variations in the use of electron transport, ranging from the oxidation of water to the oxidation of organic compounds such as pyruvate or formate for the generation of a proton gradient. There are yet other uses made of electron fluxes such as the reduction of NAD and NADP involved in biosynthetic processes and detoxification. My limited familiarity with the P-450 systems precluded a critical evaluation of the massive data on these important reactions. What other uses does nature make of electron transport? What are some of the electron transport enzymes doing in the plasma membrane and in some organelles such as chromaffin granules? Are they participating in transport of ions? Are they involved in regulations of the membrane potential that has such a powerful effect on ion fluxes through channels? There is so much to be explored!

III. Bacteriorhodopsin

A. Reconstitution with Denatured Bacteriorhodopsin and with Proteolytic Fragments

Bacteriorhodopsin, present in the membrane of *Halobacterium halobium,* is probably the best-characterized transport protein. On illumination of the bacteria it catalyzes the outward movement of protons (Oesterhelt and Stoeckenius, 1973). Bacteriorhodopsin contains 248 amino acids and has an M_r of 27,000. It can be reconstituted into liposomes by a variety of procedures (Racker, 1973; Racker and Stoeckenius, 1974; Racker *et al.,* 1979). Its amino acid sequence has been determined (Ovchinnikov *et al.*, 1979; Khorana *et al.*, 1979) and confirmed by the sequencing of the gene. Based on electron diffraction methods, it was proposed that bacteriorhodopsin spans the membrane seven times, forming a continuum of α helices (Henderson and Unwin, 1975). The carboxyl end faces the cytoplasmic side and contains about 20 amino acids in the aqueous phase (Gerber *et al.*, 1977). The amino terminus is on the outside of the cell.

In a remarkable series of papers, Khorana and his collaborators have used methods of reconstitutions, proteolytic digestion, as well as genetic, chemical, and immunological analyses to elucidate the structure and mode of action of bacteriorhodopsin. I shall describe some of the information gained from reconstitution experiments. First, complete delipidation was achieved by starting with ^{32}P-labeled purple membranes that were suspended in Triton X-100 and eluted from BioGel A 0.5 M in the presence of 0.25% deoxycholate. Solubilization of bacteriorhodopsin with SDS resulted in loss of ability to bind retinal and partial loss of secondary structure. Renaturation of delipidated bacteriorhodopsin after denaturation by SDS occurred by addition of retinal and soybean lipids in the presence of cholate (Huang *et al.*, 1981). A most remarkable reconstitution was achieved with chymotryptic fragments of bacteriorhodopsin (Liao *et al.*, 1983). Two fragments, C-1 containing amino acids 72–248 and C-2 containing amino acids 1–71, were recombined in the presence of retinal at pH 6.0. Vesicles formed from this tertiary complex pumped protons at a rate close to that achieved with native bacteriorhodopsin.

B. Phospholipid Specificity

Delipidated bacteriorhodopsin was reconstituted with a variety of phospholipids by the cholate-dialysis procedure. Endogenous polar phospholipids from *H. halobium* were particularly effective (Lind *et al.*,1981; Höjeberg *et al.*, 1982). A mixture of phosphatidylglycerolphosphate and glycolipid sulfate at a ratio similar to that in the natural membrane gave maximal rates of proton pumping. Marked stimulation of proton pumping by valinomycin was observed when the natural lipids were used for reconstitution. Among the single lipids phosphatidylglycerol yielded vesicles with the highest pumping rates as well as with the largest stimulation by valinomycin. The various head groups of the individual lipids had characteristic effects both on pumping and on vesicle formation by reconstitution.

In spite of the use of endogenous phospholipids from *H. halobium* in various mixtures, these investigators have not succeeded in imitating the natural right side out configuration of bacteriorhodopsin. This is in agreement with our observations. In our hands the procedure described by Happe *et al.* (1977) to obtain right side out vesicles at a low pH of reconstitution has yielded, at best, vesicles that excrete protons at a marginal rate. This failure was at least partly due to the observed bleaching of bacteriorhodopsin. With the new procedures of renaturation described by Khorana, this obstacle could be overcome.

The second unnatural feature is that a large ratio of phospholipid to protein ($40:1$ w/w) is required to obtain maximal rates of proton translocation. This is characteristic of virtually all reconstitutions with highly purified proteins, although in some instances mentioned earlier (Ca^{2+} pump and cytochrome oxidase) this ratio can be reduced to $3:1$ or $5:1$. This is still far removed from the natural ratio of about $1:2$. It appears from our limited experience that vesicles reconstituted in the right side out configuration are active at a lower lipid-to-protein ratio than inside out vesicles.

Lind *et al.* (1981) reported that valinomycin stimulated H^+ transport rates by about 5-fold and the extent of proton translocation by about 10-fold. Using endogenous lipids, but not soybean lipids, under optimal conditions, more than 15-fold stimulation by valinomycin was observed. It seems likely, as proposed by the authors, that this phenomenon is related to a greater degree of proton impermeability of the vesicles reconstituted with endogenous phospholipids. The effect of valinomycin was most pronounced when there was Na^+ outside and K^+ inside the vesicles. This is consistent with the proposition that valinomycin K^+ acts as a counterion. But it is apparent from the data and from our own observation that valinomycin invariably also accelerates the efflux of H^+ in the dark (even when the extent of H^+ translocation is similar with or without valinomycin). In any case, the use of chemically and genetically modified bacteriorhodopsin in reconstitutions with pure lipids should help to elucidate the mechanism of this most primitive proton pump that does not involve an electron transport chain.

C. Reconstitution with Retinal Analogs

Bacteriorhodopsin contains one retinal molecule bound as a Schiff base to the lysine 216 amino group (Bayley *et al.*, 1981a). Substitution of the retinal by a variety of analogs has provided new insights into its mode of action and has eliminated some of the proposed mechanisms for H^+ pumping. It was shown by Oesterhelt and Christoffel (1976) that 1,3-dehydroretinal replaced retinal in its biological function but 5,6-epoxiretinal did not. Marcus *et al.* (1977) replaced retinal with several analogs and concluded that considerable alterations in chemical structure, particularly in the ring of retinal, can be tolerated with retention of some functional activity in spite of the fact that the action spectrum was considerably shifted. The most illuminating replacement was achieved with phenylretinal by Bayley *et al.* (1981b). Since full function was retained after reconstitution, proton abstraction from the β-ionone ring of retinal, which had been proposed as part of the mechanism during proton translocation, was ruled out. Since the planar phenylretinal was more rapidly bound to the apomembrane, this observation suggested that the binding of retinal may be predicated on a planar molecular configuration.

Lesson 27: On the simplicity of nature (and its investigators)

I called bacteriorhodopsin the most "primitive" H^+ pump. I first wrote "simple" but erased it. Have a look at the photocycle of bacteriorhodopsin and try not to be confused. "Primitive" refers to a stage in evolution with a lesser risk of being proven wrong. According to Whitehead, we should seek simplicity and then distrust it. Surely bacteriorhodopsin is not simple. Is it the most likely among the pumps to give up its secrets? Perhaps so, but we haven't as yet even scraped the surface of its simplicity. Look at it the other way. Shall we be able to segregate physically the steps of the photocycle? Perhaps we may after all be better off to dissect less primitive pumps that have evolved to an elevated level of simplicity.

IV. Halorhodopsin and a Bacterial Na⁺ Pump

Even though they do not pump protons, these two transporters are described here because they are ion motive force generators.

A. The Light-Driven Cl⁻ Pump of Halobacterium halobium

This is a cousin of the bacteriorhodopsin proton pump and also contains retinal. It catalyzes a photocycle that is accompanied by Cl^- import (Schobert and Lanyi, 1982). The 20,000-dalton polypeptide containing the chromophore has recently been reconstituted into liposomes and also incorporated into planar bilayers, where it catalyzes an electrogenic movement of Cl^- (Bamberg *et al.*, 1984; Bogomolni *et al.*, 1984).

B. A Bacterial Na⁺ Pump

A remarkable variant in bioenergetics is a sodium pump present in *Klebsiella aerogenes* and in some other fermenting microorganisms driven by the decarboxylation of oxaloacetate or of other carboxylic acids. The enzyme (molecular weight 68,000) contains biotin and is firmly anchored in the cytoplasmic membrane. The protein was incorporated into soybean phosphatidylcholine liposomes by octylglucoside dilution and was shown to pump Na^+. Optimal reconstitution was achieve at 2.7% octylglucoside, and virtually no Na^+ transport was observed below 1.5% (Dimroth, 1981). $^{22}Na^+$ uptake, measured after passage of the vesicles through Dowex-50, was completely abolished by avidin or nigericin. In the obligate anaerobic bacteria *Proprionigenum modestum,* the decarboxylation of methylmalonyl-CoA establishes an Na^+ gradient, which in turn generates ATP via an Na^+-ATPase (Dimroth and Hilpert, 1984). A remarkable achievement for a modest microorganism.

Lecture 9

Facilitated Diffusion, Symporters, and Antiporters

This lecture must start with an apology superimposed on that in the Preface. Perhaps more than any other field this one is in a stage of rapid developments and fluxes of both solutes and concepts. The gulf between studies of native membranes and intact cells or organelles on the one hand, and of reconstituted vesicles on the other, is much greater than it could be. I have chosen examples that demonstrate the kind of new information that has been obtained by reconstitution experiments.

I. Transporters of Plasma Membranes

Transport of ions and nutrients takes place through a variety of mechanisms. Even the same nutrient, e.g., glucose or an amino acid, may be transported by either facilitated diffusion, Na^+ symport, or H^+ symport, depending on the cell type. Indeed in one and the same cell, multiple transporters may be used, e.g., for the transport of alanine or methionine. We have very little insight into the physiological reasons for this variety of options. An obvious possibility is that these nutrients vary in availability for different cells. For example, the red blood cell is exposed to a comfortable and stable concentration of glucose. It may therefore be content with a simple transport system of facilitated diffusion. On the other hand, intestinal cells living in competition with microorganisms, or microorganisms struggling in competition with other microorganisms have developed more sophisticated and effective import mechanisms.

A. The Glucose Transporter of Human Erythrocytes and Yeast

The basic properties of the transporters that facilitate the diffusion of D-glucose into erythrocytes of many mammalian cells have been reviewed (Jones and Nickson, 1981; Wheeler and Hinkle, 1982, 1985). It is the consensus of most (though not all) investigators who have studied this transporter that it is a protein with an M_r of about 55,000 and that it appears in SDS–PAGE as a broad band in zone 4.5. Some of the controversies have arisen from the use of affinity labels. It was reported that maltosylisothiocyanate irreversibly inhibited the transporter and almost exclusively labeled a protein with an M_r of 100,000 (Mullins and Langdon, 1980). Other investigators using various radioactive affinity labels in the presence and absence of glucose or cytochalasin B have reported labeled bands in SDS–PAGE of 50,000, 60,000, 90,000, 180,000, and 200,000 (cf. Wheeler and Hinkle, 1982), pointing to the ambiguity of experiments with affinity labels that are not absolutely specific. Unfortunately, too few specific reagents are available. I have therefore emphasized the usefulness of the reconstitution assay that allows us to measure the functional activity during purification of a transporter.

The experiments pioneered by Kasahara and Hinkle (1976) pointed to the 55,000-dalton protein band as representing the D-glucose transporter of human erythrocytes. Early reconstitution experiments yielded proteoliposomes with low transport activities, but an improved purification and the use of a better assay procedure gave rates that corresponded to 10–40% of the rate expected from kinetic studies of intact cells and that were five times faster than observed with membrane vesicles from erythrocytes (Wheeler and Hinkle, 1982). A useful table summarizing various isolation procedures and reconstitutions into liposomes and bilayer lipid membranes is included in the review by Jones and Nickson (1981).

Radiation inactivation studies have indicated a molecular weight of 180,000, corresponding to a tetramer of the transporter in the red cell membrane (Cuppoletti et al., 1981). The reconstituted vesicles revealed, by freeze fracture electron microscopy, particles of 60 Å in diameter, which would be consistent with a dimer of the 55,000-

dalton polypeptide (Hinkle *et al.*, 1979). Taking the number obtained by the treacherous radiation inactivation measurements at its face value, this difference in assembly may contribute to the lower rates of transport seen in the proteoliposomes. Only very limited attempts have been made to explore assembly with well-defined phospholipids. A study with the glucose transporter of adipocyte plasma membrane reconstituted with egg PC and PE (1:1) showed that a fluid membrane favored operation of the glucose transporter and appears to rule out the operation of a rigid channel (Melchior and Czech, 1979).

The freeze–thaw reconstitution procedure discovered during investigation of the D-glucose transporter because of the malfunction of a refrigerator has been used by most investigators for the study of this transporter, but other methods, e.g., detergent removal, are also effective. Hess and Andrews (1977) devised a novel method for the isolation of actively transporting vesicles that could be separated from nonfunctional vesicles on a density gradient. Goldin and Rhoden (1978) used the same principle of a "transport specificity fractionation" approach for the identification and purification of the glucose transporter of erythrocytes. This method consists of reconstituting membrane proteins extracted with cholate into liposomes, followed by removal of cholate by hollow-fiber dialysis. Then, by exploiting a physical property of the reconstituted vesicles that depends on transport activity, they are separated from inactive vesicles. More specifically, only vesicles with a functioning glucose transporter are filled with glucose. Their higher density permits a separation on an isoosmotic density gradient. These innovative experiments confirmed the conclusion that the "4.5 band" of erythrocytes represents the glucose transporter.

Early failures to demonstrate inhibition of glucose transport by hydrophobic compounds that effectively inhibited the transport in erythrocytes were shown to be due to binding to excess phospholipids (Wheeler and Hinkle, 1982).

Lesson 28: On inhibitors of membrane functions

We need to pay attention to the chemistry of the inhibitors when we explore their effect on proteins located in membranes. If we are dealing with hydrophilic inhibitors, they act—in most instances—only from one side of the membrane. When an inhibitor like *N*-ethylmaleimide added from the outside affects the plasma membrane on the cytosolic side, we should suspect that there is a transporter that facilitates the transgression of the inhibitor.

When a hydrophobic inhibitor that is effective in native membranes is ineffective in reconstituted systems, there are two obvious possibilities. One is that there was something wrong with the reconstitution. The second is that the inhibitor was drowned in a bath of excess phospholipids, often required for reconstitution. We should therefore express concentrations of hydrophobic inhibitors in terms of milligram membrane lipids and proteins rather than in terms of their concentration in water.

I have said this before, but I'll have to keep saying it until editors, even of prestigious journals, are willing to accept it.

Mullins and Langdon (1980) and Shelton and Langdon (1983) raised the possibility that the 55,000-dalton protein is a proteolytic product of a larger anion transporter present in band 3. In response to this serious possibility, two laboratories showed independently, with antibodies against the transporter, that the glucose transporter of fresh erythrocytes, fractionated by a procedure that avoided proteolytic degradation, traveled as a 55,000-dalton polypeptide (Baldwin and Lienhard, 1980; Sogin and Hinkle, 1980). Moreover, it should be pointed out that Shelton and Langdon (1983) measured only D-glucose transport since they used glucose oxidase for the assay in the reconstituted vesicles. Thus no controls with L-glucose were recorded. In view of the observations of van Hoogevest *et al.* (1983) that vesicles reconstituted with purified glycophorin facilitated a transport of L-glucose that was sensitive to 100 μM DIDS, experiments that do not exclude such nonspecific hexose passage must be viewed with caution. Phloretin, the inhibitor used by Shelton and Langdon, is as unspecific as DIDS.

It is apparent from this controversy and our studies on the Na^+/Ca^{2+} antiporter mentioned earlier that transport activity in reconstituted vesicles may be misleading if not accompanied by careful evaluation of its specificity (e.g., stereospecificity) or susceptibility to physiological regulation. Reconstitution of purified band 4.5 into large unilamellar liposomes by reverse-phase evaporation resulted in rapid hexose transport (5 μmol $\times$ min^{-1} mg protein^{-1}) that was sensitive to cytocholasin B (Carruthers and Melchior 1984). In contrast to freeze–thaw sonication this procedure also supported some cytocholasin B-sensitive transport by reconstituted band 3, but only at a rate of 5–9% of that supported by band 4.5. These findings confirm some of the observations of Shelton and Langdon (1983) but do not answer the question of whether the activity is catalyzed by contamination of band 3 with band 4.5 or whether there is a larger, active precursor of band 4.5. The data reemphasize the point made earlier that more than one reconstitution procedure should be explored.

A new development is the solubilization and reconstitution of a transporter for D-glucose from yeast plasma membranes that appears to be similar to the erythrocyte transporter in catalyzing a stereospecific facilitated diffusion of D-glucose (Franzusoff and Cirillo, 1983). This is an interesting beginning with many future possibilities of nutritional and genetic manipulations.

Evidence has been presented that in fat cells, stimulation of glucose uptake by insulin is caused by the movement of glucose transporters from intracellular Golgi-rich membranes to the cell surface (Suzuki and Kono, 1980; Cushman and Wardzala, 1980). Alternatively, it was proposed that insulin increases the glucose transport activity of both microsomal and plasma membrane fraction (Carter-Su and Czech, 1980). A recent paper (Kono et al., 1982) contains convincing data of six- to eightfold stimulation by insulin of glucose transport in reconstituted vesicles from plasma membranes with a parallel 50% drop in the Golgi-rich fraction. Since the majority of the transporter is in the Golgi-rich fraction, the 50% loss roughly accounts for the increase in the plasma membrane. Four agents known to mimic insulin effects, namely, trypsin, vanadate, H_2O_2, and p-chloromercuriphenyl sulfonate had a similar effect on glucose transporter distribution. With this elegant study, reconstitution of crude membrane fractions as an approach to physiological problems has come of age.

B. The Na^+/Glucose Symporter

The first direct observations on the role of Na^+ in glucose transport in the intestine were made by Riklis and Quastel (1958) who stated that "Na^+ must be present for glucose absorption to take place." Crane (1965) first clearly formulated an Na^+/glucose symport mechanism and visualized it to be catalyzed by a carrier with one site for glucose and one for Na^+. Brush border vesicles were isolated from rat small intestine and shown to catalyze a D-glucose-specific uptake which was dependent on a high concentration of Na^+ in the medium (Hopfer et al., 1973). Fairclough et al. (1979)

achieved reconstitution of the Na^+/glucose symporter with the freeze–thaw sonication procedure. Using defined lipids (phosphatidylserine and cholesterol at a molar ratio of 1 : 1), the partially purified Na^+/glucose symporter from pig kidney was reconstituted by a variant of the freeze–thaw sonication method (Koepsell et al., 1983). Before reconstitution, the crude membranes were precipitated twice after solubilization with deoxycholate, followed by removal of the detergent. Enrichment of a 52,000 MW protein in fractions with high glucose transport activity was observed. Cholesterol greatly enhanced transport in vesicles reconstituted with purified phospholipids (Ducis and Koepsell, 1983). These studies should be compared with the work in the laboratory of Crane (Malathi et al., 1980; Malathi and Preiser, 1983) who reported on the Na^+/glucose transporter from kidney cortex and intestinal brush border vesicles from rabbit. By controlled proteolysis of the membrane with subtilisin followed by solubilization with 0.4% deoxycholate, the transporter was purified over 20-fold by passage through a concanavalin A–Sepharose column. The preparation showed a single band with an M_r of 165,000 in SDS–PAGE and was reconstitutively active using a simple sonication procedure. However, these observations need further exploration since only about 5% of the activity in native vesicles was observed. Of particular interest is the observation that interaction with an antibody against the protein enhanced the activity 2-fold. I shall return to the possibility that an antibody may facilitate the proper incorporation of a protein into liposomes. Neither of the two divergent values for the M_r of the glucose transporter coincide with an estimate of 71,000 daltons, based on the interaction of a protein with fluorescein isothiocyanate that could be blocked by glucose in the presence of Na^+ (Peerce and Wright, 1984).

C. The Inorganic Anion Transporter of Erythrocytes

There is vast literature on the properties of this transporter which plays a crucial role in the gas exchange processes that take place in erythrocytes. Extensive reviews (Cabantchik et al., 1978; Lowe and Lambert, 1983) deal with its kinetic properties, its architecture, its sensitivity to inhibitors, its mode of action and physiological role in gas exchange, and its control of intracellular pH and cell volume. Although inorganic anion transporters are present in other cells, I shall restrict the discussion to the erythrocyte system because the characterization of the transporter from other sources is often based on studies with inhibitors that are not as specific as advertised (e.g., DIDS). Moreover, erythrocytes have been chosen for reconstitution studies for a very good reason, namely, the abundance of the transport protein (ca. 20% of the total membrane proteins) in these cells. I call it the inorganic anion transporter because of the presence of an organic anion transporter in erythrocytes with very different properties (Dubinsky and Racker, 1978).

Before discussing the isolation and purification of "band 3" according to the nomenclature of Steck (Fairbanks et al., 1971), the 90,000-to 100,000-dalton protein that contains the inorganic anion transporter, it should be mentioned that there are several other minor components in this band that have similar migration properties in one-dimensional SDS–PAGE but can be separated by more sophisticated methods. Among them is the α subunit of the Na^+, K^+-ATPase. Properties, such as binding of some glycolytic enzymes (Strapazon and Steck, 1976) and phosphorylation of tyrosine residues (Dekowski et al., 1983), may serve as additional identification probes, provided they can be shown to be specific for the transport protein and not shared by proteins that migrate similarly. Indeed, reconstitution studies may shed light on the possible physiological significance of interactions between these associated proteins and the transporter.

Purification of band 3 has been achieved either by negative purification, i.e., extraction of less hydrophobic proteins by detergents, or by differential extraction with Triton X-100 (cf. Cabantchik et al., 1978). Negative purification by extraction of erythrocyte ghosts by sonication in the presence of cholate, followed by reconstitution of the insoluble proteins with phospholipids by the cholate dialysis procedure was used in early reconstitution experiments (Gasko et al., 1976). In these reconstituted vesicles transport of $^{32}P_i$, which is a convenient assay with slower transport kinetics than uptake of chloride, was virtually completely blocked by SITS or phloretin. Prolonged extraction with Triton X-100 solubilized the protein, which was purified by column chromatography and found to be reconstitutively active (Lukacovic et al., 1981). Extraction of ghosts with Triton X-100 yielded a protein of about 85% purity. Phospholipids solubilized in Triton X-100 were added and Bio-Beads SM-2 were used to remove excess Triton. This reconstitution procedure was used to evaluate the role of phospholipids (Köhne et al., 1983). PC (or PC + PE) plus cholesterol yielded vesicles that transported $^{35}SO_4^{2-}$ and the process was partially inhibited by 2,2-dinitrostilbene-4,4'-disulfonate (a reversible relative of DIDS). Addition of sphingomyelin or PS to PC during reconstitution inhibited the transport capacity of the vesicles. In an interesting paper, van Hoogevest *et al.* (1983) reported the reconstitution of band 3 by the Triton-removal procedure and also observed only partial inhibition of transport at 100 μM DIDS. More disturbing are the observations that (a) the vesicles also catalyzed L-glucose transport, and (b) both $^{35}SO_4^{2-}$ and L-glucose were transported by vesicles that were reconstituted with purified glycophorin instead of band 3. Moreover, DIDS at 100 μM (and even at 10 μM) inhibited the $^{35}SO_4^{2-}$ transport and also inhibited the influx of L-glucose. These experiments not only cast doubts on the value of using inhibitors such as DIDS without establishing their specificity, but raise questions about studies on D-glucose transport that are not controlled by experiments with L-glucose, as pointed out earlier. Indeed, stereospecificity is usually a better criterion than sensitivity to an inhibitor. Moreover, this is a valuable study, since it points to deficiencies in other reconstitution experiments that do not include controls with vesicles reconstituted with an unrelated protein such as glycophorin.

A more sophisticated and more tedious variant of the Triton X-100 reconstitution procedure was described using octylglucoside (which by itself is a poor solubilizer of the transporter) to remove residual Triton (Darmon *et al.*, 1983). A fluorescent anion was used to measure fluxes, and a 50% inhibition by disulfonic stilbenes was interpreted to be caused by random orientation of the transporter in the proteoliposomes. Treatment of the protein with DIDS prior to reconstitution yielded 70% inhibition. In view of the above mentioned studies on DIDS-sensitive ion fluxes in glycophorin proteoliposomes, it seems that more specific inhibitors of this transporter are badly needed. We have recently reconstituted the DIDS-sensitive inorganic anion transporter from human erythrocytes by a simple procedure of freeze–thawing. Transport rates far in excess of those observed with intact erythrocytes were observed (I. Ducis, M. A. Kandrach, and E. Racker, unpublished experiments).

Lesson 29:   On the use of inhibitors

Only uninhibited investigators use inhibitors (Carl Cori). Since inhibited investigators make little progress, Dr. Cori and the rest of us continued to use inhibitors—sometimes with caution, sometimes without. There are certain rules to remember. Young inhibitors are always specific; the older they get the less specific they become. Sometimes we can increase the specificity of inhibitors by demonstrating protection by specific substrates (remember that substrates may be lacking specificity as well). The specificity of inhibitors may be increased by control of the pH, ionic strengths, membrane potential, and presence of cofactors. For example, at a neutral pH in the presence of NAD (but not in its absence), iodoacetate is by three orders of magnitude more inhibitory than iodoacetamide for glyceraldehyde-3-P-dehydrogenase. In contrast, NAD actually protects the same enzyme against inhibition by NEM. The specificity of an inhibitor may be enhanced by the use of a second, reversible inhibitor with an overlapping but different specificity. For example, DIDS (which often "dids" more than advertised) reacts with SH groups. To eliminate this mode of action, SH groups can be protected with DTNB, that can subsequently be removed as in the case of the proton pump of clathrin-coated vesicles. Perhaps the most important rule is: if you have to use inhibitors, use more than one, preferably ten.

There is another inorganic anion transporter in secretory epithelial cells that is distinct from the transporter in erythrocytes and has become of considerable interest in view of the defectiveness of chloride transport in sweat gland cells of patients with cystic fibrosis (Quinton, 1983). Respiratory epithelial cells appear to have a similar Cl^- transporter (Widdicombe and Welsh, 1980; Langridge-Smith et al., 1984) that is also impaired in cystic fibrosis (Widdicombe and Welch, 1985). With the successful reconstitution of the solubilized Cl^- transporter from tracheal apical membranes (W. P. Dubinsky, unpublished experiments) progress in the understanding of this defective transport system in cystic fibrosis patients should be forthcoming.

D. The Nucleoside Transporter

The transport of nucleosides into cells has been extensively reviewed (Plagemann and Wohlhueter, 1980; Young and Jarvis, 1983). The properties of this transporter are of particular interest in view of the use of nucleoside analogs against virus infections and tumors (Suhadolnik, 1979). Differences have been reported between the properties of the widely studied transporter in erythrocytes and that of tumor cells (Slaughter et al., 1981; Wohlhueter and Plagemann, 1982). Lack of inhibition by nitrobenzylthioinosine (6-[4-nitrobenzylthiol]-9β-ribofuranosyl purine) (NBMPR) in some lines of cultured tumor cells (Slaughter et al., 1981; Wohlhueter et al., 1978;

Belt, 1983) has opened up chemotherapeutic potentials for this inhibitor in conjunction with anticancer nucleoside analogs. More basic studies are needed, however, before a rational use of the inhibitor as a protector of normal cells can be used. Such studies are being conducted, and considerable information on the properties of the transporter in human erythrocytes is available (Wu *et al.*, 1983; Young and Jarvis, 1983).

The mechanism of the nucleoside transporter appears to be similar to that of the transporter that catalyzes the facilitated diffusion of glucose discussed earlier. The purification and reconstitution of the nucleoside transporter has been achieved by the same procedure Kasahara and Hinkle (1976) had used for the glucose transporter (Tse *et al.*, 1984; Belt *et al.*, 1984). The same (broad) 4.5-kDa polypeptide was active in transporting uridine and glucose in reconstituted vesicles. However, in contrast to the uptake of glucose, the transport of uridine was sensitive to NBMPR and insensitive to cytocholasin B (Jarvis *et al.*, 1982). The availability of photoreactive NBMPR (Wu *et al.*, 1983) has allowed for the demonstration of an excellent correlation between function and labeling of the 4.5-kDa band. Moreover, pig erythrocytes, which lack a functional glucose transporter but transport nucleoside rapidly, show a 4.5-kDa band that reacts covalently with photoactive NBMPR; the nucleoside transport-deficient sheep cell does not. It therefore appears that the 4.5-kDa band of human erythrocytes is a mixture of two or more proteins. The observations that NBMPR binding sites per cell doubled in HeLa cells between the G and S phase of the cell cycle (Cass *et al.*, 1979) and that ConA increased uridine uptake by two- to threefold in lymphocytes point to an important phenomenon of regulation. Studies on the internalization of the nucleoside transporter similar to those performed with the glucose transporter may be rewarding.

E. The Lactate/H^+ Transporter

The export of lactate is an essential feature of all glycolysing cells. In bacteria, erythrocytes, and tumor cells lactate is excreted together with a proton (Harold and Levin, 1974; Halestrap, 1976; Spencer and Lehninger, 1976; Dubinsky and Racker, 1978). In Ehrlich ascites tumor cells this is the major pathway for elimination of intracellular protons (Stone *et al.*, 1984a). A second proton exporter is an ATP-driven proton pump similar to that in clathrin-coated vesicles, which I have discussed earlier. In these cells the

Na^+/H^+ antiporter represents a third, but minor, mechanism of proton excretion.

The importance of having an insight into the mechanism of a transporter before embarking on reconstitution is illustrated by the story of the lactate and pyruvate transporter of erythrocytes. It had been shown that pyruvate transport into sealed erythrocyte ghosts was inhibited by SITS and other inhibitors of the chloride–bicarbonate transporter (Rice and Steck, 1976), suggesting "that pyruvate shares a common transport with chloride." These observations were confirmed by Dubinsky and Racker (1978), but only under conditions when the chloride transporter generated a proton gradient that facilitated the movement of lactate. When the ionic composition was reversed so that the proton gradient counteracted lactate flow, addition of SITS actually stimulated lactate uptake into the erythrocytes. Indeed, to demonstrate lactate dependent proton uptake, it was necessary to block the inorganic ion transporter with SITS. Thus, a design of a reconstitution experiment must take into consideration the possibility of the presence of another transporter which may interfere with the detection of the first. Reconstitution of the lactate transporter from Ehrlich ascites tumor cells has been achieved but the data have been variable; reconstitution with the transporter from rabbit erythrocytes looks more promising (M. Newman, S. Nisco, and E. Racker, unpublished experiments, 1984). One of the complications is that the leakage of lactate from small, unilamellar vesicles takes place at an appreciable rate and is increased when nonspecific hydrophobic proteins are incorporated into the liposomes. These observations resemble those made on studies of the glucose transporter discussed earlier. Thus, it became essential to use a specific inhibitor. Those used earlier, e.g., mersalyl and others, were unsuitable because they rendered the vesicles even more leaky to lactate. A synthetic anhydride between lactic and formic acid was shown to be a suitable inhibitor of lactate transport in erythrocytes and EAT cells (Johnson et al., 1980). Unilamellar giant liposomes prepared by the freeze–thaw procedure without sonication have large trapping volumes and slow leakage of lactate of pH 8.0 (M. Newman, S. Nisco, and E. Racker, unpublished experiments, 1984). They are currently being used for the incorporation of the lactate transporter from rabbit erythrocytes. An important observation (Jennings and Adams-Lackey, 1982) is the labeling in rabbit erythrocytes of a polypeptide of an M_r of 43,000 by radiolabeled reduced DIDS at 10^{-4} M, a concentration that is much higher than required for the labeling of the inorganic anion transporter, but is in line with concentrations required to inhibit lactate transport. The authors realize that these data do not prove that the 43-KDa band is in fact the lactate transport protein. However, they point out that in human erythrocytes which have a V_{max} for lactate transport 100-fold lower than rabbit erythrocytes, there was little or no labeling by reduced DIDS in the 43-kDa region. Fractionation of the 43-kDa protein of rabbit erythrocytes away from other hydrophobic interfering proteins may be a valuable step in future attempts to reconstitute the transporter.

F. Na^+-Linked Transporters

1. Amino Acid Transporters

In pioneering work Christensen (1975) has elucidated several amino acid transport systems in mammalian cells. One of these, the Na^+-driven system A which transports certain neutral amino acids (e.g., alanine, methionine, serine, and proline), has been resolved and reconstituted. There are several aspects of system A that are particularly interesting biologically. I shall discuss them, since they are amenable to future studies in reconstituted systems. Most amino acid transporters in mammalian cells, including system A, show broad specificity, i.e., they allow passage of several amino acids. Moreover, a single amino acid can be transported by more than one transporter. In view of this complexity, it is fortunate that α (methylamino) isobutyric acid (MeAIB) has been shown to serve as a relatively specific substrate for system A (Christensen *et al.*, 1965).

Of the eight amino acid transport systems that have been described in hepatocytes, system A stands out as being regulated by hormones such as insulin or glucagon (Kilberg, 1982). Adaptive regulations of system A, e.g., by starvation or by hormones, have been demonstrated in many other mammalian cells (Guidotti *et al.*, 1978). Of particular interest is the observation that the AIB transport catalyzed by plasma membrane vesicles isolated from 3T3 cells is inhibited by cAMP-dependent protein kinase (Nilsen-Hamilton and Hamilton, 1979).

I have become very intrigued by the findings of an increased transport of AIB or MeAIB in transformed cells (Foster and Pardee, 1969; Isselbacher, 1972; Boerner and Saier, 1982). Methionine, a substrate for system A as well as for system L that transports leucine, appears to have an altered metabolism in tumor cells (Stern *et al.*, 1984; Hoffman, 1984). This is manifested by observations that, compared to normal cells, a variety of human tumor cell lines have low levels of intracellular-free methionine and S-adenosylmethionine and elevated levels of S-adenosylhomocysteine when exposed to a medium containing homocysteine but lacking methionine. A higher dependency on external methionine also appears to be a common feature of transformed and human malignant cells. These remarkable and apparently contradictory properties of tumor cells will be discussed further in Lectures 11 and 12.

Reconstitution experiments with system A of amino acid transport have been reported from the laboratories of Isselbacher, Johnstone, and Oxender. Nishino et al. (1980) have described the reconstitution of an Na⁺-stimulated and Na⁺-specific transporter from mouse fibroblast BALB/c 3T3 cells transformed with Simian virus 40. Peripheral proteins were removed by treatment with dimethylmaleic anhydride according to Shanahan and Czech (1977) and the residue was extracted with 2% sodium cholate. The crude extract was reconstituted with PE/PC at a lipid-to-protein ratio of 10 at a final concentration of 2% cholate, which was removed by passing the mixture through a Sephadex G-50 column. The vesicles in the emerging void volume were centrifuged, the pellet resuspended, frozen, thawed, and sonicated for 20 sec. The reconstituted vesicles had transport properties similar to those of native vesicles, but some distinct differences were noted as well. As pointed out by the authors, further purification of the receptors is needed to characterize the components reponsible for transport activity. No data were reported comparing reconstituted vesicles from normal and transformed cells. Since insulin receptor is now available in pure form, co-reconstitution with system A transporter may shed light on the mode of its activation by insulin.

Johnstone and Bardin (1976), Johnson and Johnstone (1982), and McCormick et al. (1984) have reconstituted the system A transporter from Ehrlich ascites tumor cells. A three-to fourfold stimulation of AIB uptake by Na⁺ was observed with vesicles reconstituted by dialysis of a cholate/urea extract of plasma membranes. Cecchini et al. (1979) also reported reconstitution of the system A transporter from Ehrlich ascites tumor cells. Although it is clear that several laboratories have achieved the reconstitution of this system, it appears that progress has been slow, partly due to a great variability of results (multiple personal communications). M. Schenerman, M. S. Kilberg, and E. Racker (unpublished experiments) observed that inclusion of a detergent during freeze–thaw reconstitution of EAT membranes yielded very reproducible proteoliposomes that catalyzed Na^+-dependent uptake of alanine or AIB. A more extensive exploration of the phospholipid requirement, both during purification and reconstitution and of regulatory mechanisms (e.g., phosphorylation–dephosphorylation), may serve to allow more rapid progress in this important area of transport research.

2. Other Na^+-Linked Transporters

In addition to the glucose and amino acid transporter discussed earlier, there are numerous other Na^+-, or H^+-linked transport systems in intact cells that have not as yet been isolated and reconstituted. In some instances the first step, isolation of active plasma membrane vesicles, has been accomplished. Na^+-stimulated transport of P_i was first observed in Ehrlich ascites tumor cells (Scholnick *et al.*, 1973) and in vesicles obtained from the same source (Racker *et al.*, 1980a). The transporter from these vesicles has been reconstituted into liposome vesicles (J. Belt, unpublished observation). Vesicles from mouse fibroblasts (Hamilton and Nilsen-Hamilton, 1978; Lever, 1977) and renal and intestinal border membranes (Cheng and Sacktor, 1981) exhibit a similar Na^+-stimulated uptake of P_i.

A biologically fascinating transport system is the inward movement of Cl^- that is linked to Na^+ or to $Na^+ + K^+$ influx. This system has been extensively studied in erythrocytes, kidney, and Ehrlich ascites tumor cells and was recently reviewed in an article by multiple authors from various laboratories (Warnock *et al.*, 1984). In view of the implications of hormonal control and possible control by phosphorylation (Alper *et al.*, 1980) it would be important to reconstitute this transport system. Initial attempts have thus far not met with success (R. Huganir, personal communication).

A useful summary table for a variety of Na^+ gradient-dependent transporters in vesicles from renal proximal tubules involving amino acids, sugars, ions, and organic metabolites, such as lactate and Krebs cyclic intermediates, can be found in an article by Sacktor (1982).

G. Cation Antiporters of Plasma Membranes

1. The Na^+/H^+ Antiporter

This antiporter plays a pivotal role in pH homeostasis in both eukaryotic cells (Roos and Boron, 1981; Gillies, 1981; Nuccitelli and Deamer, 1982) and microorganisms (Zilberstein *et al.*, 1982; Guffanti *et al.*, 1981; Krulwich, 1983). In bacteria, decisive experiments have been possible because of the availability of mutants that lack this transport system. The proposed role of the antiporter in mitogenesis will be discussed in Lecture 11. An important role for the Na^+/H^+ antiporter was included in a stimulating hypothesis by Skulachev (1979). He proposes that the Na^+/K^+ gradient plays the same role as a $\Delta \mu_{H^+}$ buffer in bacteria as creatine phosphate does in the maintenance of the ATP reservoir in some eukaryotic cells. Very complex regulations of various ion channels have been invoked for this mechanism to function.

Lesson 30: What we can learn from studies of transport mechanisms in the plasma membrane

Life as we know it is dependent on the continuous import of wanted ions and nutrients and on the export of unwanted ions and waste products. ATP-driven pumps are preferentially used for transport processes that require an active movement against a gradient, but numerous other mechanisms have been devised by nature. All these activities take place without the aid of a transport union. They are regulated by a marvelous network of feedback mechanisms, cooperations between solutes that are moved by symporters and antiporters. They are controlled by the electrical potential of the membrane, by phosphorylation–dephosphorylation, and by hormonal activators and inhibitors. Perhaps one day in the future men and women of vision will apply some of these principles of checks and balances to human society and do away with exploitations by either employers or unions.

In mammalian cells, pH homeostasis is very complex with the participation of multiple transporters and regulatory mechanisms, some of which are cAMP-dependent and presumably involve phosphorylation and the availability of ATP (Roos and Boron, 1981). There is virtually nothing known about the relative quantitative contributions of the various H^+ extrusion mechanisms, except that they differ from cell to cell. In the proximal tubule of the kidney (Warnock and Rector, 1981) the Na^+/H^+ antiporter predominates; in Ehrlich ascites tumor cells the lactate/H^+ transporter is by far the major H^+ excretor, while the Na^+/H^+ antiporter plays a minor role (Stone *et al.*, 1984a). In erythrocytes the inorganic anion transporter is likely to be a major contributor to proton fluxes, whereas in the medullary collecting duct of the kidney and in the bladder of some species, an electrogenic, ATP-driven proton pump is responsible for proton secretion (Stone *et al.*, 1983b). The Na^+/H^+ transporter has been studied in dog kidney cells in tissue culture and its possible participation in cell differentiation opens up an interesting approach to the problem (Rindler and Saier, 1981). A remarkable Na^+/H^+ antiporter was described by Heefner and Harold (1982), who observed ATP-stimulated Na^+/H^+ fluxes in vesicles from *S. faecalis* when proteolysis was inhibited by addition of protease inhibitors during preparation. I shall return in Lecture 11 to these observations and to the proposed role of the Na^+/H^+ antiporter in growth stimulations.

Reconstitution into Liposomes of Amiloride-Sensitive Na^+ Transport. Reconstitution of the Na^+/H^+ transporter from rabbit kidney vesicles, after solubilization of the membrane with octylglucoside, has been described (LaBelle and Lee, 1982). The concentration of octylglucoside used to solubilize the protein from microsomes was high (3%) and a ratio of phospholipid to protein of about 7:1 was used. The uptake of Na^+ was inhibited by the diuretic amiloride, when present at high concentrations (1.8 mM for 50% inhibition), and by external cations. Ca^{2+} inhibited significantly at 1 mM; K^+ at 5 mM (LaBelle, 1984). The octylglucoside extract was found to be stable and should therefore lend itself to further purification and reconstitution. Future comparisons between the properties of the transporter in the reconstituted vesicles and the Na^+/H^+ antiporter of intact cells should establish whether or not the proteoliposomes are representative of the physiological transporter.

2. The Ca^{2+}/Na^+ Antiporter

Extensive reviews have been written on the operation of the electrogenic Ca^{2+}/Na^+ antiporter in intact cells (Blaustein, 1976; Mullins, 1976; Baker, 1976). Active vesicles have been isolated from plasma membranes of the heart of rabbits (Reeves and Sutko, 1979) and dogs (Pitts, 1979). Solubilization, partial purification, and reconstitution of the transporter from bovine heart vesicles have been described (Miyamoto and Racker, 1980). In these experiments the Na^+-dependent uptake of Ca^+ was shown to be inhibited by external Na^+. If the transporter was extracted with Triton X-100 rather than with cholate, it did not exhibit this control mechanism characteristic of native vesicles. Great progress has been made in the purification by exposing the membranes to a protease which removed "impurities" without impairing the reconstitution activity of the transporter (Wakabayashi and Ghoshima, 1982). It will be important to establish, once the transporter is obtained in pure form, whether the protein is intact or was nicked during protease treatment. It should also be pointed out that the flux measurements in reconstituted vesicles are in the opposite direction to those occuring under physiological conditions, simply because it is easier to load liposomes with Na^+ than with Ca^+, which interferes with reconstitution. Moreover, further experiments are needed to establish the role of asymmetry in this transporter. At present we do not even know whether the reconstituted vesicles are inside out or right side out or mixtures of both.

A recent report of the Na^+/Ca^{2+} antiporter purification and reconstitution from brain synaptic plasma membranes looks most promising. By using the transport-specific fractionation procedure, which I described earlier, an over 100-fold purification was achieved. The activity coincided with the enrichment of a 70,000-dalton protein (Barzilai et al., 1984).

3. The Ca^{2+}/H^+ Antiporter

The operation of a Ca^{2+}/H^+ antiporter in Ehrlich ascites tumor cells has been described (Hinnen et al., 1979). The process is electrogenic, as indicated by stimulation or inhibition by valinomycin, depending on the presence of Na^+ or K^+ in the external medium. Ca^+ entry was sensitive to μM concentrations of ruthenium red. Reconstitutions of this transporter into liposomes was not as reproducible as the reconstitution of this transporter from retinal disk of the outer rod segments of the retina (Racker et al., 1980b). The latter activity is of particular interest as a possible step in the visual process. The transporter was reconstituted in the presence of a large excess of soybean phospholipids fortified with a neutral lipid such as Q_{10}. Either cholate dilution or sonication was used for the reconstitution of the membrane fragments into liposomes. The method of measurement was similar to that used for the mitochondrial Ca^{2+}/H^+ antiporter that will be discussed later.

H. Transporters in Bacteria

It is remarkable how relatively little work on reconstitution of bacterial transport systems has been published, in spite of all the obvious advantages bacteria offer and the information already available from genetic analyses (Wilson, 1978; Slayman, 1978). There are obvious obstacles. Growing kilogram quantities of microorganisms that are needed for extensive biochemical studies is more laborious than obtaining slaughterhouse material or making a phone call to Pelfreeze. Commercially available bacteria, unless grown to specifications, are often poor starting materials. Moreover, bacteria may have multiple transport systems for a single nutrient. For example, *E. coli* has at least five systems that transport galactose (Wilson, 1978) and has even more diverse mechanisms for the transport of amino acids than some mammalian cells. Until recently, many attempts at reconstitution of bacterial transporters have failed. The exceptions are the proton pumps discussed earlier, e.g., F_1F_0 and bacteriorhodopsin, the PTS (Postma and Roseman, 1976; Dills *et al.*, 1980), the alanine transporter of thermophilic bacteria PS3 (Hirata *et al.*, 1984), the gal P transporter of *E. coli* (Henderson *et al.*, 1983) and the lactose transporter of *E. coli* (Newman and Wilson, 1980). However, the list of successful reconstitutions of bacterial transporters is growing rapidly. Some were discussed in the last lecture; I shall mention a few more in this one.

1. The Lactose Transporter of *E. coli*

I have selected the lactose transporter for detailed discussion because it illustrates the wealth of information that can be obtained from reconstitution experiments. The lactose transport system, induced by the presence of lactose during growth of *E. coli,* was genetically analyzed in 1955 (Cohen and Rickenberg, 1955). It took 10 years before Fox and Kennedy (1965) isolated the "M protein" that was implicated in the process. These investigators treated *E. coli* membranes with NEM in the presence of a high-affinity galactoside substrate. After washing away the substrate and excess NEM, radioactive NEM was used to label the transporter. I would like to emphasize the usefulness of this method, which is used more often by immunologists than by biochemists. For example, DTNB, a reversible reactant with SH groups, was used to protect the proton transporter of clathrin-coated vesicles against an irreversible inhibitor (Xie *et al.*, 1983).

An important advance in the approach to the lactose transport problem was the enrichment of the transporter proteins with a recombinant plasmid, thus purifying the protein without fractionation

(Teather *et al.*, 1980). It was not until Newman and Wilson (1980) developed a method of solubilization and reconstitution with octylglucoside that decisive biochemical experiments with a pure transport protein could be performed (Kaback, 1983). In these reconstituted vesicles either lactose counterflow or lactose accumulation was measured. Since lactose uptake takes place by an electrogenic lactose/H^+ symport mechanism, the membrane potential was collapsed with K^+-valinomycin to allow the substrate to accumulate.

The lactose transporter is an excellent example in which the intelligent use of a radioactive photoaffinity label (*p*-nitrophenyl-α-D-galactopyranoside) served as a valuable aid during the course of purification (Newman *et al.*, 1981; Foster *et al.*, 1982), which included an initial step of removal of impurities from the membrane with urea and cholate (see Lesson 5). Another important feature was inclusion of phospholipids during purification (see Lesson 7), which stabilized the protein and permitted its purification to homogeneity by DEAE–Sepharose chromatography. The final purification was 35-fold with a 50% yield based on recovery of the radioactive photoaffinity label, an achievement in line with the estimates of the content of the lac protein in the genetically amplified strain of *E. coli*. A single broad band with an M_r of about 33K was observed in SDS–PAGE, in agreement with the early measurements with the NEM-labeled carrier (Jones and Kennedy, 1969). Yet, based on the DNA sequence (Büchel *et al.*, 1980) which was matched by the amino acid composition, the protein has a molecular weight of 46,504, which is similar to the value obtained by gel permeation chromatography (König and Sandermann, 1982). This is another example of a discrepancy between the real molecular weight and the M_r observed in SDS–PAGE (see Lesson 21).

Newman *et al.* (1981) also introduced a useful and simple variant of transport activity assay for the lactose transporter, which should have general applicability when the protein to be reconstituted is present in dilute solutions. Reconstitution was achieved by octylglucoside dilution of the proteoliposomes, but instead of concentration by centrifugation, proteoliposomes were filtered onto a Millipore filter and the vacuum was released. Radioactive lactose was then applied to the filter, and after given time periods, the vacuum was reapplied in order to remove the excess radioactive lactose that remained outside the vesicles. Although it appears that much lower activities were observed with this method than by the conventional procedure, it should be admirably suited for rapid assays of reconstituted transport activities, e.g., when many fractions need to be assayed during purification (and no affinity label is available).

The method of reconstitution of the lactose transporter finally adopted consisted of octylglucoside dilution followed by freeze–thaw sonication. Electron micrographs revealed that this yielded unilamellar vesicles with diameters between 50 and 150 nm. Freeze fracture pictures showed a uniform distribution of particles 85–90 Å in diameter, suggesting that the protein was incorporated as an oligomer. How many polypeptides are involved could perhaps be more closely estimated by correcting for the increase in diameter caused by shadowing, using as standards gold particles or other particles of known size (Telford *et al.*, 1984). An alternative method of purification and reconstitution of the lactose transporter was described by Wright *et al.* (1982). Remarkably, only lactose counterflow and no net lactose accumulation was observed with these proteoliposomes. These vesicles could be very useful for an analysis of the requirements for exchange versus net uptake and might be used to test, by freeze–fracture experiments, the hypothesis (Robertson *et al.*, 1980) that the dimer is required for uptake while the monomer can catalyze exchange. The proteoliposomes prepared according to Newman *et al.* (1981) catalyzed net lactose uptake with V_{max} and K_m closely corresponding to the values obtained with native vesicles.

An interesting co-reconstitution experiment with *o*-type cytochrome oxidase from *E. coli* was performed by Matsushita *et al.* (1983). With reduced coenzyme Q_1 as electron donor, these proteoliposomes accumulated lactose against a concentration gradient, and the process was abolished by uncouplers that collapsed the $\Delta \mu_{H^+}$.

Experiments with the reconstituted lactose transporter provided persuasive evidence for a sequential mechanism of lactose transport in which the deprotonation of the transporter is rate limiting. This explains why counterflow was much faster than net uptake and why, in contrast to net uptake, the counterflow of lactose was not controlled by the pH. Accordingly, net lactose transport was strongly inhibited when D_2O instead of H_2O was used, but the exchange was not. On the other hand, when $\Delta \psi$ was used to drive lactose uptake no deuterium isotope effect was observed. It appears that under these conditions the release of proton from the transport proton was not rate-limiting. An excellent discussion of these findings was presented by Kaback (1983).

Imaginative experiments were performed with electron inactivation using reconstitution as a tool of analysis. Of particular interest is the observation that the transporter appears to be twice as large (ca. 90K) when energized by an electron donor than either without it or in the presence of an uncoupler; once again, these observations point to an oligomeric structure of the transporter operative in active transport.

I have discussed these studies in such detail because, in my view, they represent a model of how to get the most out of reconstitution experiments. I am certain there is more to come with the availability of monoclonal antibodies and the use of proteolytic digestion of right side out and inside out vesicles. I have discussed experiments of this type in the lecture dealing with bacteriorhodopsin. Analogous experiments with the lactose transporter are still puzzling (Kaback, 1983) and need to be extended.

The reconstitution of lactose transport has shown that a single polypeptide with a molecular weight of 46,000 (and an M_r of 33,000) catalyzes an active transport under appropriate conditions. The experiments with the reconstituted vesicles have lent strong support to the chemiosmotic mechanism proposed by Mitchell. Moreover, they have yielded novel information on the mechanism of action and the structure of the transporter that could not have been readily obtained with native vesicles or intact cells. They will provide new information in the future with the use of antibodies and proteases, experiments already in progress. I hope that the future will also include experiments that allow or disallow us to refer to the protein as a "carrier." Meanwhile, I have preferred to use "transporter," a name with less stigma and controversy.

2. The Phosphoenolpyruvate/Sugar Phosphotransferase System (PTS)

The PTS was first discovered by Kundig *et al.* (1964). The earlier work has been extensively reviewed (Postma and Roseman, 1976; Dills *et al.*, 1980). Curiously, this highly evolved and complex system of sugar transport in prokaryotes has been abandoned during evolution for the much simpler mechanisms of sugar transport in higher organisms discussed earlier. How often do we encounter such a trend from complexity to simplicity in evolution? In a remarkable series of 12 consecutive papers covering over 100 pages published through the generosity of the *Journal of Biological Chem-*

istry, Saul Roseman and his collaborators (1982) deal with the latest developments in PTS in his laboratories. The transfer of the phosphoryl group from PEP to enzyme I, HPr, and then to one of the sugar-specific proteins, EIIA^{Man} (an integral membrane protein) or EIII^{Glc} (a soluble or peripheral protein) are described in detail. Two other integral membrane proteins, EIIB^{Man} and EIIB^{Glc}, participate in the phosphorylation and transport of mannose and glucose across the membrane. The isolation and properties of the soluble enzymes (EI, crystalline HPr, and EIII^{Glc}) are described in detail, and some genetic and kinetic properties of these soluble systems are also dealt with. In the last paper, experiments with vesicles are described but there is nothing about reconstitution of the membranous protein.

Meanwhile, other investigators have attended to the problem of vectorial group translocation of sugar by reconstitution experiments. In fact, in the preceding issue of the *Journal of Biological Chemistry* (Erni *et al.,* 1982) the purification and reconstitution of the glucose-specific EII enzyme of the PTS of *Salmonella typhimurium* was described. Solubilization was achieved with polydisperse octyloligooxyethylene (octyl POE), a useful detergent with a critical micellar concentration of about 6mM and therefore, like octylglucoside, readily removed by dialysis. After solubilization, the glucose-specific EII enzyme was purified by isoelectrofocusing and chromatofocusing as the major steps, obtaining an almost 40-fold purification compared to the activity in the extract. A functional assay was used, coupled to the other component of PTS (Enzyme I, HPr, and enzyme III^{Glc}) with sonicated phosphatidylglycerol as the source of phospholipid. In SDS–PAGE the protein has an M_r of about 40,000, while the reconstitutively active protein exists as a complex of 105,000 MW.

Homogenous mannitol EII from *E. coli* was reconstituted into liposomes by the octylglucoside dilution procedure (Leonard and Saier, 1983). The optimal concentration of octylglucoside was 1.2%, similar to that required for the reconstitution of the lactose transporter (Newman and Wilson, 1980) and of bacteriorhodopsin (Racker *et al.,* 1979). The mannitol-1-P/mannitol transphosphorylation was strongly inhibited at 1.5% or higher concentrations of octylglucoside. It should be noted here that some mammalian transport systems (e.g., the Na$^+$,K$^+$ pump) require 2% or even higher concentrations for reconstitution (Racker *et al.,* 1979). The rate of radioactive mannitol-1-P formation inside the proteoliposome was dependent on enzyme II concentration—in fact, it showed a sigmoidal response, suggesting that an oligomer form was required for the vectorial transport.

3. Other Transport Systems in Bacteria

Among the numerous carbohydrate transport systems in bacteria (Dills *et al.*, 1980), only a few have been studied by reconstitution. The melibiose/Na^+ symporter has been solubilized and reconstituted (Tsuchiya *et al.*, 1982) by the methods used for the lactose/H^+ symporter. It resembles the lactose/H^+ symporter in some respects but not in others (Cohn and Kaback, 1980). Future studies of vesicles reconstituted with the pure transporter should reveal more intimate differences and shed light on the role of cation specificity. A Ca^{2+}/H^+ antiporter from *Mycobacterium phlei* has been solubilized with cholate, partially purified and reconstituted by sonication with *M. phlei* phospholipids (S. H. Lee *et al.*, 1979). The corresponding activity from *Azotobacter vinelandii*, together with a membrane potential-driven Ca^{2+} uptake present in this bacterium, has also been reconstituted (Zimniak and Barnes, 1983).

It would be of considerable value to obtain reconstitution of the so-called shock-sensitive systems responsible for the transport of carbohydrates as well as amino acids (Berger and Heppel, 1974). Regrettably, no convincing reconstitution experiments have been reported with any of the substrate-binding proteins (cf. Wilson, 1978), although some of them are clearly involved in the transport process and in chemotaxis.

II. Transporters of Mitochondria

A. Ca^{2+}, Na^+, and K^+ Transport Systems of the Inner Mitochondrial Membrane

There is general agreement that mitochondria have separate transporters responsible for the influx and efflux of Ca^{2+}. There is also almost unanimous agreement that the influx is an electrogenic import system, but there is no agreement on the mechanism of efflux (Nicholls and Akerman, 1982; Carafoli, 1982). In fact, it seems likely in view of the multitude of propositions, including active and passive mechanisms and H^+ and Na^+ antiporters, that a clear understanding will not be reached until more specific inhibitors are available for each of the ion fluxes that take place in mitochondria. Then reconstitution experiments can be carried out with purified mitochondrial transporters from a variety of tissues, establishing the contribution of each to the complex interplay of ion fluxes.

1. The Ca^{2+} influx into Mitochondria

In the past, contradicting claims about Ca^{2+} influx into mitochondria were at least partly due to differences in experimental design. These differences were summarized in an excellent review by Nicholls and Åkerman (1982). In earlier years emphasis was placed on massive Ca^{2+} influx, under artificial conditions of high external Ca^{2+}, that established the capability of mitochondria to accumulate micromolar amounts of $Ca^{2+} \times mg^{-1}$ of protein. In later years the experimental design became more and more physiological, emphasizing steady-state conditions and the interplay with Ca^{2+} efflux. As is so often the case, by moving to more physiological conditions, the data become more difficult to interpret kinetically because other ion fluxes greatly influence the results. In any case it is apparent that the driving force for Ca^{2+} accumulation is the Ca^{2+} electrochemical potential that includes the membrane potential as well as the extra- and intramitochondrial Ca^{2+} concentration. The latter is greatly influenced by the presence of P_i, which in turn is transported by mechanisms to be discussed later. Two moderately specific inhibitors are available to inhibit Ca^{2+} influx into mitochondria: ruthenium red (Moore, 1971) and low concentrations of La^{3+} (Mela, 1968). It is thus possible to measure in the presence of ruthenium red the efflux of Ca^{2+}. Numerous observations of hormonal activation of Ca^{2+} influxes have been reported, e.g., with α-adrenergically activated mitochondria from rat heart (Kessar and Crompton, 1981) and rat liver (Taylor et al., 1980). It will be of considerable interest to repeat many of these experiments in the future in reconstituted systems with purified transporters as the rate-limiting factor. However, thus far experiments with purified "transporters" and on reconstitutions have been too ambiguous to justify their acceptance as being representative of the uniporter of Ca^{2+}. Nevertheless, in each case further studies seem warranted. An antibody against a mitochondrial surface glycoprotein (42 kDa) was shown to inhibit Ca^{2+} uptake as well as NADH oxidation–induced efflux (Panfili *et al.*, 1980). Since the glycoprotein is peripheral, it is more likely to act as a "binding protein," facilitating influx rather than acting as a transporter. Jeng and Shamoo (1980) have isolated a 3000-dalton polypeptide capable of ruthenium- and La^{3+}-sensitive Ca^{2+} transport into an organic solvent, but more convincing reconstitutions into liposomes need to be done. An 8800-dalton polypeptide facilitating Ca^{2+} uptake has been separated from cytochrome oxidase preparations and reconstituted into liposomes (Saltzgaber-Müller *et al.*, 1980). The authors made no claim, however, that this protein represents the natural uniporter because of several discrepancies, including lack of inhibition of Ca^{2+} transport by ruthenium red.

2. Ca^{2+} Efflux from Mitochondria

It is apparent that facilitated efflux of Ca^{2+} takes place in mitochondria. It can be distinguished from the ruthenium red-sensitive uniporter acting in reverse, e.g., after collapsing the $\Delta \psi$ by uncouplers. Active Ca^{2+} efflux, which can be studied in the presence of ruthenium red, involves either Na^+ or H^+ counterflow. Rat liver mitochondria catalyze an electrically neutral Ca^{2+}/H^+ antiport (Puskin *et al.*, 1976), whereas mitochondria from heart and other tissues exhibit an Na^+-dependent efflux of Ca^{2+} (Crompton *et al.*, 1978). However, the same mitochondria also catalyze an Na^+-independent Ca^{2+} efflux (Crompton *et al.*, 1977). Early studies from our laboratory demonstrated an ATP-dependent uptake of Ca^{2+} into inside out, submitochondrial particles from bovine heart (Loyter *et al.*, 1969; Christiansen *et al.*, 1969). Since there is no ATP-driven Ca^{2+} pump in mitochondria, these experiments are consistent with a Ca^{2+}/H^+ antiporter mechanism, with ATP serving to generate a proton gradient. Similarly, we have pointed out previously (Dubinsky *et al.*, 1979) that it is difficult to decide in intact mitochondria whether the well-studied Na^+ stimulation of Ca^{2+} efflux (Carafoli, 1982) is catalyzed by an Na^+/Ca^{2+} antiporter, as in the case of plasma membranes or by the sequential action of two transporters, an Na^+/H^+ and a Ca^{2+}/H^+ antiporter. We have not been able to demonstrate by reconstitution experiments the presence of a Ca^{2+}/H^+ antiporter in bovine heart mitochondria using a procedure that was successful for the reconstitution of the Ca^{2+}/H^+ antiporter present in plasma membranes of bovine heart or brain (Miyamoto and Racker, 1980). Obviously negative experiments, particularly involving reconstitutions are not persuasive. On the other hand, both the Na^+/H^+ and Ca^{2+}/H^+ antiporters from bovine heart mitochondria have been successfully reconstituted into liposomes (Dubinsky et al., 1979). A few unusual features that emerged from these studies deserve emphasis, since they may be applicable to other transport systems. First, the Ca^{2+}/H^+ antiporter was extracted from submitochondrial particles of bovine heart with Triton X-100 in the presence of La^{3+}. This cation enhanced the solubilization of the Ca^{2+}/H^+ antiporter by Triton. Removal of Triton (>99%) was achieved by ammonium sulfate precipitation of the protein in the presence of cholate and phospholipids (Hoffman, 1979). From the precipitate, the Na^+/H^+ antiporter was preferentially extracted with $C_{12}E_8$ and the Ca^{2+}/H^+ antiporter was extracted from the $C_{12}E_8$-insoluble residue with 2.5% lysolecithin. These simple procedures resulted in between 5- to 10-fold purification of both transporters. A 30-fold purification of the Ca^{2+}/H^+ transporter was achieved on a sucrose density gradient, which effectively separated it from the Na^+/H^+ antiporter. A most remarkable property of the Ca^{2+} transporter was its heat-stability (15 min at 75°C) in the presence of lysolecithin. In the absence of lysolecithin the activity decayed even at 30°C, so that lysolecithin had to be added to demonstrate sensitivity of the transporter to trypsin at this temperature.

3. K^+/H^+ antiporter

An interesting K^+/H^+ antiporter has been described in rat liver mitochondria. The observations that it is regulated by Mg^{2+} and inhibited by quinone (Nakashima and Garlid, 1982) should make it a challenging system for reconstitution experiments.

B. The Adenine Nucleotide Transporter

The properties of this transporter, which catalyzes a one-to-one exchange of ADP and ATP across the inner mitochondrial membrane, have been extensively reviewed (Klingenberg, 1976; Vignais, 1976). Its role in oxidative phosphorylation as an importer of ADP and exporter of ATP is fully recognized. What is not widely appreciated is that this transporter plays a vital role in cells with mitochondria that are lacking both respiration and the ability to catalyze oxidative phosphorylation. It was shown that when import of glycolytic ATP into these mitochondria was blocked by bongkrekate (Šubíc et al., 1974) or by op 1 mutation (Kováč et al., 1967), yeast cells lost the ability to grow and to transport precursor proteins into the mitochondria (Nelson and Schatz, 1979a). What are the vital functions of mitochondria in respiration-defective mutants that require import of ATP via the nucleotide transporter? In elucidating the function of the adenine nucleotide transporter, two inhibitors have played a major role: atractyloside (Bruni et al., 1962) and bongkrekate (Henderson and Lardy, 1970). In addition to the experiments mentioned above, they have played a useful role as photoaffinity labels, in kinetic experiments, and in the determination of asymmetry in mitochondria, submitochondrial particles, and reconstituted vesicles, because the hydrophilic atractyloside is a nonpermeable, whereas bongkrekate is a permeable inhibitor. Moreover, they have aided in the isolation of pure transporter. Although inactive, the carboxyatractyloside complex of the transporter (Brandolin et al., 1974) is more stable in Triton than the unligated protein. Thus it was possible to isolate native proteins from several sources. The protein which is abundant in bovine heart mitochondria purified as the inhibitor complex migrates in SDS–PAGE with an M_r of 30,000. In the absence of SDS, it was determined to be a dimer with an M_r of about 60,000 (Hackenberg and Klingenberg, 1980).

A live and reconstitutively active adenine nucleotide transporter was purified from submitochondrial particles, after removal of F_1 and other surface protein, by extraction with cholate and fractionation with ammonium sulfate (Shertzer and Racker, 1974, 1976). Cruder preparations isolated from SMP (Racker, 1962) were quite stable at $-80°C$, whereas the highly purified preparations from stripped particles lost considerable activity after freezing. In SDS–PAGE gels it migrated as a major band with an M_r of about 30,000. The purified transporter, after reconstitution into liposomes by sonication, catalyzed ADP/ATP exchange at a rate of 300–400 nmoles $\times$ min^{-1} $\times$ mg protein^{-1}. Similar rates were reported in vesicles reconstituted with Triton X-100–solubilized transporter (Krämer and Klingenberg, 1979). Optimal reconstitution was probably not obtained by the various sonication methods used in both laboratories, and other procedures should be explored. Although rates are lower than expected of a pure protein, the reconstitution experiments of Shertzer and Racker (1976) allowed for the following conclusions: The purified transporter catalyzes in liposomes an electrogenic exchange. Valinomycin, in the presence of K^+, stimulated ATP influx/ADP efflux but inhibited ADP influx/ATP efflux. The lipid composition that yielded the highest activity was PE/PC/cardiolipin at a ratio of $4:1:1$. In contrast to inside out submitochondrial particles, which were sensitive to bongkrekate but not atractyloside, the reconstituted vesicles were sensitive to both inhibitors, suggesting that all of them were right side out. Finally, it should be pointed out that reconstitution could be achieved with crude inner mitochondrial membrane preparations; this served as an assay during the purification of a reconstitutively active protein.

C. The Phosphate Transporter

The history of the discovery of this transporter has been reviewed elsewhere (Banerjee and Racker, 1979). The transporter catalyzes a neutral P_i/OH exchange and is required for supplying P_i to the matrix of mitochondria during oxidative phosphorylation (Chappell, 1968). In attempts to reconstitute this transporter, a protein was extracted with octylglucoside from bovine heart mitochondria and partially purified (Banerjee and Racker, 1979). Incorporation of this protein into soybean phospholipids gave rise to transport of P_i, which was stimulated over three-fold by valinomycin and nigericin in the presence of K^+. The rate was about 1000 nmoles $\times$ min^{-1} $\times$ mg protein^{-1}. With P_i inside the vesicle, a similar rate of P_i/P_i exchange was noted which was not influenced by the presence of ionophores. Optimal reconstitution was obtained by 7 min sonication of PE/PC/CL at a ratio of $4:1:1$. P_i transport was sensitive to mersalyl and NEM. The inhibition by the latter was partially prevented by the presence of P_i, as in the case of mitochondria. Interaction of radioactive NEM with a 30,000-dalton protein was also inhibited by P_i. Reconstitution of the P_i transporter, together with mitochondrial ATPase and nucleotide transporter, yielded vesicles that catalyzed ATP hydrolysis that was partially inhibited by bongkrekate or atractyloside. In spite of the kinetic resemblance of this right side out reconstituted system with mitochondria, the observed electrogenicity of this P_i transporter is contrary to the widely accepted view that in mitochondria the process is electrically silent. It is possible that in mitochondria the electrogenicity of P_i transport is obscured by compensating ionic fluxes, or alternatively, that we have purified a protein responsible for a different (uniport) P_i transport, perhaps related to that observed in rat liver submitochondrial particles (Wehrle *et al.,* 1978). Whatever the final explanation for these experiments will be, they represent the first attempts (with limited success) to assemble, with partially purified proteins, an atractyloside-sensitive ATPase of mitochondrial orientation in a reconstituted system.

P_i transport with a protein from bovine heart mitochondria, partially purified by an alternative procedure, was reconstituted by freeze–thaw sonication (Wohlrab, 1980; Wohlrab and Flowers, 1982). The optimal lipid composition for reconstitution—PE/PC/PA (3.7 : 1 : 1) was similar to the previously described system. The proteoliposomes also catalyzed both P_i/P_i exchange as well as net P_i fluxes, which were inhibited by NEM. The P_i uptake, but not the exchange, was dependent on a proton gradient (pH 6.8 inside, pH 8.0 outside). It would be of interest to know whether this transport system also responds to valinomycin, which would distinguish it from that described previously (Banerjee and Racker, 1979).

A transporter isolated from chloroplasts that catalyzes the transport of P_i, triose phosphate and 3-phosphoglycerate has been reconstituted with crude soybean phospholipids (Flügge and Heldt, 1981; Flügge *et al.,* 1983). Of particular interest are the observations on the effect of Δ pH on the movements of the three substrates for this transporter. They help to explain why on illumination chloroplasts import P_i and 3-phosphoglycerate and export triose phosphate, thereby supplying the cytosol with reducing equivalents, ATP, and a substrate for sucrose synthesis, while the P_i in the stroma is used to generate ATP.

D. Other Transporters in Mitochondria

There are numerous transport systems in mitochondria involving amino acids, di- and tricarboxylic acids, keto acids, fatty acids, intermediates of heme synthesis, as well as proteins that are synthesized in the cytoplasm and either incorporated into the inner membrane or deposited in the matrix. There is little published on the purification and reconstitution of these systems. Recently the tricarboxylic transporter from rat liver mitochondria was partially purified after solubilization with Triton X-100 and reconstituted by freeze–thaw sonication (Stipani and Palmieri, 1983). The transport activity required specifically cardiolipin. An inhibition by Triton X-100 that was reversed by cardiolipin was observed (Stipani *et al.,* 1984).

Of particular interest is the membrane potential-dependent import of many precursor proteins that are incorporated into the membranes of mitochondria or assembled in the matrix (Nelson and Schatz, 1979b; Neupert and Schatz, 1981). Attempts to identify and reconstitute the receptors for these precursor proteins are being made in several laboratories.

There is a description of the purification from pig heart mitochondria of a proteolipid that binds glutamate. It has been incorporated into liposomes that were loaded with radioactive aspartate. Addition of glutamate resulted in the release of radioactive asparate (Julliard and Gautheron, 1978). The carnitine transporter from bovine heart mitochondria has been solubilized with octylglucoside and reconstituted by sonication into soybean phospholipid vesicles (Schulz and Racker, 1979). It is noteworthy that, among many detergents tested, only octylglucoside did not inactivate the transporter. This transporter should provide a valuable tool for future studies of the poorly understood translocation of fatty acids across mitochondrial membranes.

E. Porins of Mitochondria and Bacteria

Mitochondria and outer membranes of some Gram-negative bacteria contain a representative of a group of transmembranous proteins called porins, which form voltage-modulated channels and which appear to serve as molecular sieves (Nikaido and Nakae, 1979). Porin from *E. coli* has been well characterized (Rosenbusch, 1974) and crystallized (Garavito and Rosenbusch, 1980). The protein has an M_r of 36,500 and its amino acid composition exhibits a surprisingly high polarity for a transmembranous protein with no obvious hydrophobic domain. It has been proposed (Garavito et al., 1982) that a large fraction of the polypeptide is arranged in β-pleated sheets. In view of these unusual properties of this protein, it is of interest that it has been incorporated into model lipid membranes by several investigators (Benz et al., 1978; Schindler and Rosenbusch, 1981). Channels close with increasing potential and reopen at lower voltage. Planar bilayers formed from reconstituted vesicles containing porin and a bacterial glycolipid form channels that have properties similar to those of native outer membranes. A porin isolated from *Neisseria gonorrhea* was shown to be anion selective, differing from porin of other gram-negative bacteria, but resembling the porin of mitochondria (Young et al., 1983). Mitochondrial porin has been isolated from mitochondria of plants (Zahlman et al., 1980), *Neuros-*

pora (Freitag *et al.,* 1982), and rat liver (Columbini, 1980; Lindén and Gellerfors, 1983). It allows passage of polysaccharides up to 8 kDa via a channel that appears to be much larger than that formed by *E. coli* porin (600 daltons). Reconstitution methods into planar bilayers with porin from plants and *E. coli* were described by Zahlman *et al.* (1980) and Schindler and Rosenbusch (1981).

III. Transporters of Other Organelles

A. Sarcoplasmic Reticulum

1. P_i Transport

The exact mechanisms for the storage, and particularly the release, of Ca^{2+} from sarcoplasmic reticulum have remained elusive (cf. Miyamoto and Racker, 1981). Although calsequestrin and other Ca^{2+}-binding proteins are localized inside of the SR, the amount of Ca^{2+} that can be stored in the absence of oxalate of P_i is small. P_i seems to be a likely candidate as a natural co-ion in the process of Ca^{2+} uptake and storage. It was shown (Carley and Racker, 1982) that during uptake of Ca^{2+} by rabbit SR, about one P_i was taken up for each Ca^{2+} transported. P_i uptake required external Mg^{2+}, Ca^{2+}, and ATP. A Ca^{2+} gradient was not required since in the presence of A23187, which collapsed the Ca^{2+} gradient, P_i uptake was only slightly impaired. Vesicles reconstituted with crude SR membranes had the same properties of P_i uptake, but vesicles reconstituted with purified Ca^{2+}-ATPase did not. These vesicles only stored Ca^{2+} when preloaded with either P_i or oxalate and were incapable of ATP-dependent P_i uptake. Without internal P_i neither Ca^{2+} nor P_i was taken up, but when co-reconstituted with an 0.8% cholate extract of SR, both activities were restored. Deoxycholate, which is used to extract the Ca^{2+}-ATPase, destroyed the P_i transporter. Further purification of this protein was hampered by its instability.

2. K-Channels

A most interesting study has emerged with a reconstituted K-selective and voltage-gated channel of SR (Coronado and Miller, 1980; Miller, 1982). This channel was studied by incorporating SR vesicles into planar phospholipid bilayers by osmotic membrane fusion. Currents through a single channel were measured under a variety of conditions. By using organic cations of various sizes, it was estimated that the narrowest cross section is >20 Å^2 and that a "mouth" of at least 50 Å is present on the side opposite to the side at which the SR was added (trans-side). A series of n-alkylbis-α,ω-trimethylammonium channel blockers of various length were synthesized and added to the trans-side. From these studies it was concluded that 65% of the voltage drop within the channel occurs at a distance of 6–7 Å where the restriction of the channel diameter occurs. It appears that the short-chain blockers bind in an extended-chain conformation and the long-chain blockers in a bent-over conformation, with both charges deeply inside the channel. This original approach of probing inside a channel should be applicable to the study of other voltage-gated channels, such as the Ca^{2+} channel of excitable tissues, and its response to different Ca^{2+}-channel blockers, though few channels are as "simple" as the SR K-channel. Its physiological significance is still obscure.

B. Clathrin-Coated Vesicles and Secretory Granules

In addition to the electrogenic ATP-driven proton pump there are other transporters in clathrin-coated vesicles that require study. There is the electrogenic Cl$^-$ transporter that can be measured independent of the proton pump by collapsing the membrane potential with valinomycin and K$^+$ (Xie $et\ al.$, 1983). It is much more labile than the proton pump, but attempts to solubilize and reconstitute the responsible protein are in progress (D. K. Stone, personal communication).

Highly purified clathrin-coated vesicles contain Na$^+$,K$^+$-ATPase activity. A small but reproducible stimulation of the proton pump activity be vanadate was observed under conditions that favored Na$^+$ pumping (intravesicular K$^+$). This suggests that the Na$^+$,K$^+$, and H$^+$ pumps reside in the same vesicle (D. K. Stone, X.-S. Xie, and E. Racker, unpublished experiments). Could the presence of the Na$^+$,K$^+$ pump be of some physiological significance, e.g., by participating in osmotic swelling and fusion to form endosomes?

The transport of iron into the cytoplasm after dissociation from the transferrin–transferrin receptor complex at the acid pH in clathrin-coated vesicles needs experimental exploration. The export of toxins from the acidic intracellular organelles remains to be elucidated.

The solubilization and reconstitution of the catecholamine transporter from chromaffin granules illustrates some of the difficulties encountered in reconstitution. Solubilization of the protein was obtained with 0.9% cholate in the presence of soybean phospholipids (Maron *et al.*, 1979). Removal of detergent was achieved by Sephadex filtration or dialysis. The vesicles catalyzed a reserpine-sensitive uptake of radioactive 5-hydroxyltryptamine (5-HT) against its concentration gradient, provided a Δ pH was artifically generated with nigericin in the presence of K^+ ($K_{in} > K_{out}$). Inhibition by reserpine, an effective inhibitor of amine uptake in chromaffin granules, was never as effective in proteoliposomes; the more phospholipids were present, the less effective it was. Moreover, there were appreciable blank rates in the absence of protein when a Δ pH was generated. Thus, several lines of evidence had to be presented in favor of a valid reconstitution. In addition to inhibition by reserpine, which under some conditions was found to inhibit amine uptake into protein-free liposomes (D. Rhoads, unpublished observation), it was shown that 5-HT uptake into proteoliposomes was competitively inhibited by other biogenic amines with K_i values similar to those recorded with native vesicles. Moreover, saturation kinetics were observed only with the proteoliposomes and not in liposomes. In view of the large blank values with liposomes, an optimal lipid-to-protein ratio for reconstitutions could not be established and a ratio of about 3:1 was used. Therefore, it will be necessary to purify the transporter further before critical questions such as the role of the membrane potential can be answered without ambiguity.

Chromaffin granules, like clathrin-coated vesicles, cotransport Cl^- together with H^+. Similarly, the H^+ pump was inhibited when duramycin was added to inhibit Cl^- transport (N. Nelson, personal communication).

The 5-HT transporter has also been reconstituted from platelets. Plasma membrane vesicles were treated with sodium cholate and (without solubilization) incorporated into soybean phospholipid vesicles (Rudnick and Nelson, 1978). For the uptake of serotonin, both reconstituted and native vesicles required Na^+ and Cl^- on the outside and K^+ on the inside and were inhibited by 1 μM imipramine. Membrane vesicles, derived from dense granules of platelets, catalyzed an ATP-dependent uptake of serotonin that was inhibited by reserpine, resembling the process catalyzed by adrenal chromaffin granules (Fishkes and Rudnick, 1982). Since the transporter appears to tolerate digitonin (Talvenheimo and Rudnick, 1980) purification and reconstitution may be feasible.

Reconstitution of a solubilized preparation of the GABA transporter from rat brain was described (Kanner, 1978). The protein was extracted with Triton X-100 and fractionated with ammonium sulfate. Cholate dialysis was used for reconstitution. Like in native vesicles, the uptake of radioactive GABA into proteoliposomes was completely dependent on both Na^+ and Cl^-. Since K^+ (inside) plus valinomycin did not substitute for Cl^- it seems that this anion has a more specific function. K^+-valinomycin did, however, stimulate in the presence of external NaCl, suggesting that the process is electrogenic. An Na^+ gradient was also required; nigericin inhibited strongly. Since the proteoliposomes appear to resemble the native vesicles and liposomes do not give large blank values, this system appears very suitable for future studies with more highly purified transporters to elucidate the role of Na^+ and Cl^-.

A glutamate transporter from rat brain synaptosomes has been reconstituted with crude soybean phospholipids by the cholate dialysis procedure (Kanner and Sharon, 1978). The uptake was specific for L-glutamate; the D-isomer did not inhibit. As with native vesicles, the process was dependent on Na^+. Valinomycin stimulated and nigericin inhibited. In contrast to the γ-aminobutyrate transporter, external Cl^- was not essential, but internal K^+ was.

I started this lecture with a defensive apology. Perhaps it is appropriate to end with an offensive lesson.

Lesson 31: Passive transport, porters, channels, and facilitated confusion

I have carefully avoided the use of the expression "passive" transport, which one encounters in the literature. To call, e.g., the entry of Ca^{2+} into eukaryotic cells "passive" describes better the investigator than the process. Nature does not let hydrophilic ions enter without a passport. It invented channel, porter, and facilitated diffusion mechanisms that all require operation of specific proteins. Even uncharged organic solutes like glucose are denied unrestricted entry. Receptors, selectivity filters, gates, gating currents, and desensitization are among the devices of nature that ion gate crashers (biochemists and physiologists) study to get an insight into the regulatory mechanisms of traffic. Well-constructed phospholipid bilayers are the liquid cement protecting the interior of cells. On the other hand, the remarkable flexibility and mobility of the lipids embedded with specific proteins that control entry and exit make cells a place fit to live in.

Lecture 10

Plasma Membrane Receptors

The outcome of any serious
research can only be to make two questions grow,
where only one grew before.

Thornstein Veblen

"Receptors are proteins typically composed of several domains. They contain, by definition, at least one binding site specific for a natural ligand. Receptors interact with one or several of a variety of effector systems for which they must also possess recognition sites. A single effector system may connect with various receptors. The information for activating the effector system is entirely contained in the membrane receptor; the ligand, or specific anti-receptor antibodies, only act as triggers of receptor-mediated effects, often initiated by receptor change in conformation, micro-aggregation, redistribution, internalization or chemical modification."

This "consensus statement" summarized in a recent article (Strosberg, 1984) is well put. However, one gains the impression that all receptors delegate function to "one or several effector systems." It seems to me that most receptors believe in a "do-it-yourself" policy. For example, the EGF receptor consists of a single polypeptide chain which has a ligand binding site on one end and a protein tyrosine kinase on the other. Then there is the acetylcholine receptor, with its 5 subunits that cross the membrane, often more than once, and it also operates without effector systems. The LDL and transferrin receptors are even willing to leave their home base in the membrane to carry out their transport function, rather than delegate it to other effector proteins. Why are some receptors single polypeptide chains that cross the membrane? Why do some contain multiple transmembranous subunits? Why do some receptors travel backward and forward? Why do others, such as the sacrificial IgA receptor, leave the membrane, never to return? And, finally, why do some receptors prefer to have their mission carried out by a relay system of effectors?

The answers to these questions will come from an understanding of the functions and regulations of receptors. The complexity of the effector-dependent receptor is probably linked to the fact, cited in the quotation, that a single effector system may connect with more than one receptor and may thus allow for the remarkable regulatory mechanisms that involve multiple receptors and subtypes of receptors as well as stimulatory and inhibitory effector proteins that are interposed between the signal side on the outer side and the enzyme on the inner side of the membrane. Please notice that I speak of two sides of the membrane. About 15 years ago I wrote a chapter entitled "The Two Faces of the Inner Mitochondrial Membrane." I abandoned this more poetic nomenclature ever since I noticed that a lecture I was to give on this subject was advertised as the "Two Faeces of the Inner Mitochondrial Membrane." Although I do realize that mitochondria have not always received the esteem they deserve, I felt that this was going too far. I shall not repeat here the odorous introduction I gave to that lecture.

There is little resemblance between effector-dependent receptors and firmly anchored receptors such as the acetylcholine receptor. Obviously, serving as a channel it must span the membrane with a certain rigidity to prevent leakage, yet be capable of conformational changes required for the opening and closing of its doors to ions during its response to the agonist. The function of the LDL and transferrin receptors is also transport, but by an entirely different mechanism involving endocytosis. We have at least a beginning understanding of how they operate during their travels. The glucose receptor also wanders in and out of the plasma membrane but for a different reason, namely, for the up and down regulation of transport activity.

I have obviously overstepped the boundaries of convention by calling the glucose transporter a receptor. Of course I did it on purpose to point out ambiguities in the definition of receptors. Shall we restrict the name to ligand-responsive proteins or shall we include those who respond to an electrical signal or even to light? The light-responsive complex in the outer rod segment bears a much greater similarity to the β-adrenergic receptor than the latter to the acetylcholine receptor. The similarity between the tetrodotoxin-sensitive Na^+ channel and the acetylcholine receptor is far greater than the resemblance of either to the transferrin receptor, although all three are involved in ion transport albeit by entirely different mechanisms. Now that I have invited the Na^+ channel to take its place among the receptors, what about the numerous ion channels that respond to an electrical signal (Levitan *et al.,* 1983; Miller, 1983)? Shall we exclude the inorganic anion channel of erythrocytes because chloride is its ligand as well as its substrate? Diphtheria toxin requires Cl^- for its interaction with the membrane and its toxicity (Sandvig and Olsnes, 1984). Perhaps its site of interaction is an anion channel which we could then include in the family of receptors. How about ion channels that are controlled by a signal from within such as phosphorylation of a serine residue (Levitan *et al.,* 1983)? What is the basic difference between the EGF receptor and the cAMP-activated protein kinase? Does the receptor have to be in the plasma membrane? Are we to exclude the steroid receptors in the nucleus? What do we do with protein kinase C which responds to external stimuli but is only loosely associated with the membrane?

The major point of these introductory remarks is to emphasize that the term "receptor" has no precise definition and has come to mean different things to different investigators ever since its name was coined by Paul Ehrlich almost 80 years ago to explain the specificity of action of antibodies and drugs.

Lesson 32: Definitions of receptors

Ehrlich's "lock and key" concept of receptors did not include signals such as light or a membrane potential. Once again, with age, the concept of receptors has lost some of its preciseness. Receptors have become very fashionable and we are stuck with the name. Like generalizations, definitions aren't worth a damn if you take them too seriously. Yet we need definitions and we need classifications in biology, first to hear the motif of variations of a theme and then to get an insight into the beautiful unity as well as complexity of nature.

To get some structure into this lecture and for the sake of provocation, I propose the following classification of plasma membrane receptors.

I. RGC Receptors. They have a recognition subunit R, a G-binding protein, and a catalytic subunit. An important characteristic of these receptors is the flexibility of subunit interactions between different complexes.

II. Polypeptide Hormone and Growth Factor Receptors. They have one or more subunits of which at least one is transmembranous. They have a recognition site on the outside and a catalytic site facing the cytoplasm.

III.Channel Receptors. They have one or more transmembranous subunits which form a channel that opens or closes in response to chemical, electrical, or light signals.

IV. Transport Receptors. They respond to a ligand which is transported into the cells where it is utilized or eliminated.

V. Drug and Toxin Receptors. They are proteins or gangliosides that interact with foreign drugs and toxins. In some cases their natural ligands are known; in others, they are not.

The first two classes of receptors are basically enzymes activated by a signal; the third and fourth are transporters with an agonist or the transportee serving as the signal. It is easy to raise objections to these artificial yet useful classifications. The EGF and insulin receptors are also involved in transport, and the internalized complex may itself function as a late signal. A polypeptide receptor can be involved in transmission of a signal that alters the intracellular concentration of a secondary messenger such as inositol triphosphate. Some antibody receptors (e.g., for IgE) are associated with a channel that allows the entry of Ca^{2+}, another secondary messenger, while others such as the IgA receptor are travel receptors for the primary purpose of serving as scavengers. I shall return later to these examples.

Much progress has been made during the past few years in the purification of receptors, mainly due to the use of selective detergents and affinity chromatography with either covalently bound agonists, antagonists, or monoclonal and polyclonal antibodies. These highly purified receptors not only bind specific ligands, but more importantly, can be reconstituted into artificial planar bilayers, liposomes, or suitable acceptor cells that lack them.

I have chosen to concentrate in this lecture on plasma membrane receptors and have excluded receptors in mitochondria, nuclei, and other organelles, partly because of my lack of familiarity and partly because reconstitution attempts in this field have only just begun. For the same reasons receptors in microorganisms have been excluded, not without noting that they are a most rewarding subject of research, since they would allow for a combination of genetic and biochemical approaches to reconstitution. I have not included bacteriorhodopsin in this lecture because it seems to fit better—from a functional point of view—in the lecture on proton motive force generators. Its similarity with the rhodopsin recognition subunit on the one hand, and with channel receptors on the other, are obvious.

I. RGC Receptors

(R, recognition binding subunit; G, GTP binding subunit; C, catalytic subunit or enzyme.)

A. *The β-Adrenergic Receptors*

The resolution and reconstitution of the β-adrenergic receptors has been remarkably successful. The overall reaction is a hormone-regulated formation of cAMP, catalyzed by the adenylate cyclase (C). The stimulation or inhibition of this catalysis is mediated by guanine nucleotide binding proteins that have been called G_S (stimulatory) and G_I (inhibitory). G_S is the stimulatory protein that responds to the hormone-ligated R-protein by binding GTP, thus allowing its interaction with C to generate activated adenylate cyclase. The hydrolysis of GTP deactivates the system. Nonhydrolyzable analogs of GTP, like Gpp(NH)p, are therefore more potent stimulators. NAD-dependent ADP ribosylation of the α-subunit of G_S by cholera toxin inhibits GTPase activity of G_S, which remains in the active form. Similarly, ADP-ribosylation of G_I catalyzed by a toxin from *Bordetella pertussis* blocks the ability of hormones that inhibit adenylate cyclase and often potentiates the action of stimulatory hormones (Smigel *et al.*, 1984).

1. Purification and Reconstitution

Affinity chromatography with covalently bound ligands has played a crucial role in the purification of the R-proteins. The ligand binding of two adrenergic R proteins ($β_1$ and $β_2$) reside in proteins with an M_r of 62,000–64,000. The numerous smaller and functionally still active R-proteins that have been studied are probably products of endogenous proteinases (Stiles *et al.*, 1983).

Two crucial discoveries have facilitated the reconstitution of the R-protein. One was the fusion method developed by Schramm *et al.* (1977) mentioned earlier, which led to the first reconstitution of a partially purified R-protein and to the demonstration of interchangeability between subunits of different receptor complexes (Citri and Schramm, 1980; Brandt *et al.*, 1983). The second crucial discovery was the demonstration of GTPase activity associated with the G-protein and its marked stimulation by the hormone-ligated R-protein (Cassel and Selinger, 1976, 1977, 1978). This yielded a rapid enzyme assay without participation of the cyclase. It led to the purification of G_S (Sternweiss *et al.*, 1981) and its reconstitution together with the R-protein purified from turkey erythrocytes by affinity chromatography after solubilization with digitonin (Shorr *et al.*, 1982). The reconstituted phospholipid vesicles catalyzed a hormone-stimulated GTPase activity (Brandt *et al.*, 1983). These two proteins, reconstituted together with a partially purified preparation of C-protein, yielded phospholipid vesicles that catalyzed isoproterenol-stimulated cAMP formation in the presence of GTP (Cerione *et al.*, 1983, 1984).

The G_S-protein purified from rabbit liver has a molecular weight of about 70,000 but contains three major bands (52,000, 45,000, and 35,000). A fraction enriched in 52K (α) and 35K(β) bands was reconstitutively active (Sternweiss *et al.*, 1981). However, higher estimates of the molecular weight of the G-protein (95,000) and different M_r for both stimulatory and inhibitory subunit α (40,000–42,000) from human blood have been reported. The β-subunit of both stimulatory and inhibitory G is 35,000 daltons (Codina *et al.*, 1984).

There are a number of features in the resolution and reconstitution of the adrenergic receptors that are of special interest. It appears that, of many detergents that have been tried, only extraction with digitonin yielded a protein that binds the ligand efficiently. A clue to previous failures with other detergents is the observation that agonists (but not antagonists) protect the recognition protein from inactivation during solubilization, e.g., with deoxycholate (Nedivi and Schramm, 1984). It appears that the detergent locks the ligand into the active site even in the absence of G-protein (which normally induces the high-affinity binding of the ligand), thereby obscuring the capacity of the R-protein to bind the ligand. However, removal of the detergent releases the agonist and restores the functional activity of the solubilized recognition protein. These observations and the success story of purification of the R-protein, based on a ligand binding assay, allow us to draw two conclusions. One is elaborated in Lesson 33, the second is that digitonin, unlike deoxycholate, may stabilize the recognition protein in an active form in the absence of a ligand. It will be of interest to explore, on the one hand, whether digitonin allows stabilization of other membrane proteins that, e.g., require phospholipids for protection, and, on the other hand, whether other ligands such as acetylcholine may be locked in when present during solubilization of its receptor.

Lesson 33:   A lesson on lessons

Like other advice (solicited or unsolicited), the lessons I have quoted in these lectures should be taken with a grain of salt (or a gram of ascorbic acid). Purification of a protein based entirely on a ligand binding assay *can* yield a pure, reconstitutively active protein. Just don't count on it.

2. Mechanisms of Action of the β-Adrenergic Receptor and of its Desensitization

Experiments with highly purified and reconstituted proteins have greatly contributed to our understanding of the mechanism and of the complex interactions, for example, between stimulatory and inhibitory subunits of this receptor. The ternary complex formed between ligand, receptor, and G-protein binds GTP and thereby activates the adenylate cyclase. Deactivation is achieved by the hydrolysis of GTP by the GTPase causing the dissociation of the complex. Neufeld *et al.* (1980) and Lefkowitz *et al.* (1983) proposed that the G-protein thus serves as a shuttle between the agonist-R subunit and the cyclase enzyme and that the activation of the latter is achieved by the GTP-α (45 kDa) subunit of the G-protein.

Of particular interest is the role reconstitution experiments have played in the elucidation of the desensitization phenomenon (Lefkowitz *et al.*, 1984). There appear to be two mechanisms, one by a decrease of the receptor at the surface (down-regulation), the other by phosphorylation of a serine residue of the R-protein. It was shown by reconstitution studies that during down-regulation in frog erythrocytes, the R-protein is internalized but not degraded. When the agonist dissociates, the R-protein returns to the membrane and becomes once again a functional member of the receptor complex. In turkey erythrocytes, however, desensitization is achieved by phosphorylation of the recognition protein. The cAMPdPK has been implicated in this mechanism of desensitization.

Now that both purified stimulatory and inhibitory G as well as R proteins are available, we can look forward to interesting studies of their kinetic interactions in reconstituted systems. We can expect that, in the not too distant future, pure C protein will be available as well.

B. The Photoreceptor of the Outer Rod Segment of the Retina

The dramatic amplification after activation of rhodopsin by a single photon involves an RGC system which is remarkably similar to that of the β-adrenergic receptor (Sinozawa et al., 1979; Stryer et al., 1981). The light-activated rhodopsin interacts with a G-protein (transducin) to allow the exchange of GDP with GTP; the GTP-G protein dissociates from rhodopsin, and by a shuttle mechanism similar to that proposed for the β-adrenergic receptor complex, moves to a catalyst, a cGMP phosphodiesterase, which becomes activated. About 1000 molecules of cGMP are cleaved before the GTP–transducin complex is itself hydrolyzed to an inactive GDP–transducin.

Purified transducin contains 3 proteins with an M_r of 39,000 (α), 36,000 (β), and 10,000 (γ). The nucleotide binding site is on α, but both α and $\beta + \gamma$ subunits are required to catalyze nucleotide exchange and GTPase activity, as well as binding to rhodopsin. Kinetic analyses in reconstituted systems suggest that the $\beta + \gamma$ subunits are responsible for the GTPase activity exhibited by the complex (Fung, 1983). Published information on the method of reconstitution with egg yolk phospholipid by dialysis is rather meager. The obvious advantage of this system over the β-adrenergic receptor complex is the availability of all three components in highly purified form.

II. Polypeptide Signal Receptors

I have chosen a rather noncommittal term for this category because I could not think of a better one. Some of the receptors for polypeptide hormones and growth factors which belong in this group have been highly purified and well characterized. Some of these have specific catalytic activity, e.g., a protein tyrosine kinase. Yet it spite of large amounts of information about the structure of these proteins, we basically do not know how they function. There are several intracellular proteins that are phosphorylated by these enzymes *in vivo*, but thus far none of them have been definitely identified as a functional substrate linked to growth stimulation. There are signs that the pathway from insulin receptor to phosphorylation of ribosomal S6 protein will soon be defined by reconstitution experiments that will permit the assay and identification of the participating components. That will be only a small part of the story. How does insulin affect glycogen synthase or amino acid transport? We have to learn more about the so-called late effects of growth factors that stimulate synthesis of macromolecules. Clues to these problems are emerging from studies of three related receptors that interact with insulin and the insulin-like growth factors, IGF-I and IGF-II. The different affinities of the three growth factors to the three receptors on the one hand, and the striking variability in the distribution of the three receptors among cell types on the other hand, account for both the complexities of observations as well as the confusion caused by the divergent responses noted with different tissues. An excellent discussion of these problems was presented by Czech (1982). Observations that the pituitary hormone oxytocin exerts an insulin-like effect on rat adipocytes, but not of a mutant (Brattleboro) rat, may be useful in sorting out the various pathways of eliciting insulin-like effects (Hollenberg *et al.*, 1983).

The protein tyrosine kinase activity of the insulin receptor first described by Kasuga *et al.* (1982) appears to be stimulated by autophosphorylation at a tyrosine residue (Rosen *et al.*, 1983; Yu and Czech, 1984). Reconstitution experiments are needed to clarify some of the biological features of interactions between the insulin and IGF receptors. Of particular interest is the relationship between the observed effect of insulin, EGF, and other growth factors on the phosphorylation of protein S6 of the ribosomal 40S subunit in intact cells (Thomas *et al.*, 1982) and the *in vitro* phosphorylation of S6 by purified protein kinase C (Le Peuch *et al.*, 1983). Reconstitution studies could clarify the relationship between these phosphorylating systems.

I am including in this category some of the antibody receptors that signal important biological responses. Probably the best characterized example in this group is the IgE receptor. Like the RGC receptors, it appears to be a multisubunit complex with a recognition unit and an effector unit. The purified complex has been incorporated into liposomes (Metzger, 1983), and a partially purified multisubunit recognition protein has been incorporated by polyethylene glycol-induced membrane fusion between vesicles and rat basophilic leukemia cells, as measured by ligand-triggered release of radioactive serotonin (Estes *et al.*, 1985). An important advance in the understanding of the "effector unit" has been the isolation of a protein that binds the anti-asthmatic drug cromolyn. The latter blocks the ligand-triggered increase of Ca^{2+} influx and degranulation in some basophilic cells. The cromolyn-binding protein was incorporated into a mutant cell line with an impaired chromolyn-binding capacity by using Sendai virus envelopes as fusogenic carriers. This procedure restored the IgE-mediated response of increased Ca^{2+} influx and degranulation (Mazurek *et al.*, 1983). More recently the cromolyn-binding protein has been reconstituted into liposomes and incorporated into planar lipid bilayers together with other membrane components of rat basophilic cells. By cross-linking, IgE characteristic channel conductance was observed in the presence of Ca^{2+} (Mazurek *et al.*, 1984).

III. Channel Receptors

A. The Nicotinic Acetylcholine Receptor (AChR)

When an action potential depolarizes a cholinergic nerve terminal, acetylcholine is released. It diffuses across the synaptic cleft and binds to the AChR in the postsynaptic membrane. It is the function of the receptor to respond to this interaction with a conformational change that opens an ion channel. The resulting fluxes of K^+, Na^+, and other cations including Ca^{2+} down their electrochemical gradients depolarize the postsynaptic membrane, thereby stimulating an action potential in the postsynaptic cell. In the current wave of neurobiology, the AChR is a popular subject of research. Its involvement in the autoimmune disease of myasthenia gravis and its functional responses to antibodies and neurotoxins have broadened the significance of these studies. Biochemists, physical chemists, and electron microscopists have their field day with this complex protein. At present the AChR is the best characterized neurotransmitter receptor. In authoritative articles (Changeux, 1981; Karlin, 1980; Popot, 1983) the structure and function and the allosteric properties of this receptor have been extensively reviewed. The vast literature quoted in these reviews acknowledges the important contributions of Eldefrawi, Raftery, and many others to this field. The incorporation of the AChR into model membranes and recent information on its structure were also covered in an excellent up-to-date review by Anholt et al. (1984). I shall therefore restrict the discussion in this lecture to points of interest to biochemists who are concerned with reconstitution and the type of information we can gain from this approach.

The first point is the requirement for lipids during purification of hydrophobic proteins. As mentioned earlier, our experience with labile mitochondrial proteins led us to include phospholipids during the extraction and purification of the AChR (Epstein and Racker, 1978; Huganir and Racker, 1982) which permitted the successful and reproducible reconstitution of this protein (Anholt et al., 1984; Ochoa et al., 1983). Lipids can have two functions, stabilization and activation. Sometimes these functions can be clearly distinguished, as in the case of the H^+-ATPase of clathrin-coated vesicles, which required phosphatidylserine for both stabilization and activity (Xie et al., 1984). In contrast to the Ca^{2+}/Na^+ antiporter, which required high concentrations of phospholipids for stabilization (Miyamoto and Racker, 1980), very little phosphatidylserine was required for the H^+-ATPase. Also, in the case of the AChR, relatively low concentrations of lipids suffice for stabilization (Heidmann et al., 1980; Anholt et al., 1981). Two possibilities for the protective effect of lipids have been proposed (Heidmann et al., 1980). One involves a specific role for phospholipids as an allosteric effector and stabilizer of the protein; the second involves a direct toxic effect of the detergent, which is prevented nonspecifically by lipids. There are precedents for both interpretations. As mentioned above, the first one is clearly operative in the case of the H^+-ATPase of clathrin-coated vesicles. The second one is operative in the case of the β-adrenergic receptor, where deoxycholate induces a reversible conformational change (Nedivi and Schramm, 1984). A toxic effect of deoxycholate (but not of cholate) on cytochrome oxidase was discussed earlier. When large amounts of phospholipids are required, it seems more likely that they protect against detergent toxicity, though it is not always easy to differentiate between the two possibilities (Heidmann *et al.,* 1980).

The second point is the question of phospholipid specificity. The situation is clear when specific lipids are required for enzyme activity, as in the case of β-hydroxybutyrate dehydrogenase and H^+-ATPase of clathrin-coated vesicles discussed earlier. When specific lipids are required for activity measured after reconstitution into liposomes, as in the case of the AChR, evaluation becomes more difficult. For example, what is the role of neutral lipids that greatly stimulate the reconstitution of AChR (Kilian *et al.,* 1980) and of the Ca^{2+}/H^+ antiporter of retinal disks (Racker *et al.,* 1980b)? In elegant studies Ochoa *et al.* (1983) showed that a single phospholipid (DOPC) in conjunction with phosphatidic acid and cholesterol can be used for the reconstitution of the AChR. They showed, moreover, that part of the neutral lipid effect can be attributed to the larger size of the vesicles reconstituted with cholesterol. In addition, an effect of the lipid composition on the extent of incorporation, as well as on the orientation of the receptor, has been implicated (Anholt *et al.,* 1984). On the other hand, neutral lipids have been reported to be essential for maintenance of agonist-induced state transitions (Criado *et al.,* 1982). Without the aid of an enzyme assay, further information on the role of lipids on the AChR will have to come from physiochemical analyses of conformational changes (Changeux, 1981). With recent important advances in structure analysis (Fairclough *et al.,* 1984) new insight into the role of specific lipids should be forthcoming.

A major contribution of reconstitution experiments was to establish the functional entity of the AChR. The elimination of a role of the 43-kDa protein in transport, first shown to be a "contaminant" of AChR preparations by Neubig *et al.* (1979), was firmly established by reconstitution experiments. A third contribution was to demonstrate the functionality of monomeric AChR (Anholt *et al.,* 1980; Boheim *et al.,* 1981; Popot *et al.,* 1981; Huganir and Racker, 1980). It should be stressed, however, that such experiments, though important for basic studies, are not necessarily relevant with respect to physiological function, since assembly to dimers or polymers may contribute to regulatory or even rate functions. The fourth contribution is yet to come: reconstitution of the complex protein from individual isolated subunits. Only limited success has thus far been reported with respect to the dissociation of the β-subunit (Huganir *et al.*, 1981). In the presence of lipids, an α subunit was isolated with a K_d for α bungarotoxin of about 0.5 nM but interaction with acetylcholine could not be demonstrated (Tzartos and Changeaux, 1982) The observation that all 4 subunits must be expressed for function (Mishina *et al.*, 1984) suggests that this problem will be as difficult to solve as in the case of cytochrome oxidase. Nevertheless, the successful cloning and expression of the α subunit in yeast (Fujita *et al.*, 1985) might eventually permit the *in vitro* reconstitution of a functional receptor.

The value of antibodies in reconstitution experiments has been illustrated by the studies of Anholt *et al.* (1981). Immune precipitation of reconstituted vesicles with monoclonal antibodies indicated that receptors within a single vesicle are not scrambled but oriented either inside out or right side out. Only a very small fraction of many monoclonal antibodies tested were found to inhibit the channel function of the AChR (M. Montal, personal communication). Those that inhibit should be useful tools in the delineation of functional sites in the subunits. Those that do not inhibit should be useful in studies of reconstitution. Indeed reconstitution can be used to probe antibody–protein interaction (Huganir *et al.*, 1979). In the case of cytochrome oxidase of *Paracoccus denitrificans,* monoclonal antibodies have been used to study the role of individual subunits. Some antibodies inhibited reconstitution, others inhibited activity, yet others appeared to have no effect (E. Racker, unpublished experiments).

Reconstitution of the AChR into planar bilayers is described in great detail in the review of Anholt *et al.* (1984). The major advantage of this approach is that it permits the study of single channels and measurements of rapid kinetics of opening and closing of channels.

B. The Tetradotoxin-Sensitive Na⁺ Channel

It is well established that excitation and conduction phenomena in the nerve cell membrane are initiated by a transient and highly selective increase in Na^+ permeability. In recent years the protein, which is responsible for this phenomenon, has been purified from several sources aided by the availability of radiolabeled tetrodotoxin (TDX), and has been well characterized. The opening and closing of the channel is under the control of the membrane potential, but in the

presence of certain toxins such as veratridine, grayanotoxin or ba-
trachotoxin, the channel can be kept open until it is closed by other
toxins such as tetrodotoxin or saxitoxin. The first reconstitution of
this channel from the walking leg of lobster nerve was reported by
Villegas *et al.* (1977). Crude membrane fragments that were incapable of catalyzing Na^+ transport, were incorporated by the freeze–
thaw sonication procedure, either into crude soybean phospholipids
or into a mixture of PC, PE, PS, and PI. Attempts to purify this
protein from this source were hampered by limitations in the availability of starting material. Reconstitution of the Na^+ channel with
cholate-solubilized protein from brain has been described (Malysheva *et al.*, 1980; Goldin *et al.*, 1980). Highly purified proteins
from several sources including electric eel (Rosenberg *et al.*, 1984),
rat muscle (Weigele and Barchi, 1982; Barchi, 1983), and brain
(Hartshorne and Catterall, 1984; Tamkun *et al.*, 1984), have been
successfully reconstituted. There appear to be, however, distinct
differences between the subunit compositions of the Na^+ channel
from various sources. The brain channel is the most complex with
three subunits (260 kDa, 39 kDa, and 37 kDa), and the electric eel
the simplest with only one subunit. In reconstituted proteoliposomes, Na^+ flux was stimulated over tenfold by veratridine and was
completely blocked by TDX, if the inhibitor was present on both
sides of the membrane. Inhibition of 50% was observed when TDX
was added only to the outside surface, suggesting that reconstitution
in the right side out configuration was about 50%. Of particular
interest are studies with a scorpion toxin which did not bind to
proteoliposomes prepared with pure PC but interacted when crude
rat brain lipids were included during reconstitution. Single channels
were clearly resolved when the purified rat brain protein complex
was incorporated into planar lipid bilayers by the osmotic fusion
technique (Hartshorne *et al.*, 1985). In this system TDX blocked the
ionic current through the channel with a voltage-dependent affinity.
The ion selectivity and neurotoxin sensitivity were the same as
those observed with native vesicles. Similar experiments were described by Krueger *et al.* (1983).

C. Other Channels

The above reconstitutions were performed with proteins purified
with the aid of ligand binding assays. It appears now that in several
instances this procedure has been successful, provided digitonin or

phospholipids or both were present during protein purification. Yet we cannot take for granted that this will turn out to be a universal feature. Thus, there is need for additional functional assays. A reconstitution assay based on patch-clamping has been used for the purification of a K^+ channel from the land snail *Helix* (Levitan *et al.*, 1983, and Levitan, personal communication). Such a functional assay may be of particular usefulness when the activity of the channel is controlled by phosphorylation–dephosphorylation reactions, as is the case in this Ca^{2+}-activated and voltage-dependent K^+ channel and in the β-adrenergic RGC receptor discussed earlier.

The voltage-sensitive Ca^{2+} channel from skeletal muscle transverse tubules has been purified after solubilization with digitonin using a binding assay with the Ca^{2+} channel blocker, [³H]nitrendipine (Curtis and Catterall, 1984). This is a channel of both basic and clinical importance, in view of the wide use of Ca^{2+} channel blockers in heart diseases. It is already apparent that there is more than one mode of interaction between the blockers and the channel, as in the case with inhibitors of the AChR. I expect that reconstitution of this channel will soon be achieved and that this will allow for more penetrating studies of the mechanism of action of the large multitude of channel blockers which are being synthesized for clinical purposes.

The voltage-gated Cl^- channel of *Torpedo* electroplax was first reconstituted into phospholipids by cholate dialysis. The vesicles were enlarged to 30 μM diameter by freezeing and thawing, and single sealed bilayer patches of membranes were isolated from the surface of the liposomes with a micropipette (Tank *et al.*, 1982). This novel "liposome patch" method has obvious advantages for the study of single channels using high-resolution electrical recording methods. It should be applicable to the study of other channel proteins and perhaps even provide for a sensitive assay during purification when other, simpler methods are not available.

IV. Transport Receptors

As in the case of the polypeptide signal receptors, there is a wealth of information on the structure of receptors in this category. Moreover, their functional pathways are more clearly defined, particularly in the case of the LDL, asialoglycoprotein, and transferrin receptors. Remarkably, this was achieved without the aid of reconstitutions. I therefore refer you to recent articles written on the subject (Ashwell and Harford, 1982; Brown *et al.*, 1983; Klausner *et al.*, 1983; Daútry-Varsat *et al.*, 1983).

There are, however, aspects of these pathways that will not be solved without reconstitutions. For example, the function of the transferrin receptor is the import of Fe^{2+}. The Fe–transferrin–receptor complex is transported by endocytosis into endosomes where the acid pH dissociates the Fe^{2+} while the transferrin–receptor complex continues on its travel back to the plasma membrane. How does the Fe^{2+} escape from the endosomes? I suspect that a specific transport protein is involved in this process. How do the receptors travel back to the plasma membrane? What are the signals that direct them away from the graveyard of the lysosomes to allow them to renew their function? I have mentioned earlier some of the exciting experiments, probing by reconstitution the intracellular travel of proteins. Aided by yeast mutants, the field of endo- and exocytosis has become a goldmine for reconstitution experiments.

The asialoglycoprotein receptor functions as a scavenger, though an additional role as a signal for endosomes–lysosome fusion has also been suggested. Indeed, reconstitution experiments have been initiated to explore this possibility (Klausner *et al.*, 1980). Of particular interest is its capability to guide galactose-terminated saccharides linked to therapeutic agents to the parenchymal cells of the liver (cf. Ashwell and Harford, 1982).

Another clinically important receptor is the phosphomannosyl receptor, which plays a critical role in the uptake and intracellular travel of lysosomal glycosidases (Neufeld, 1981; Sly *et al.*, 1981).

V. Drug and Toxin Receptors

There are binding sites on membranes for drugs and foreign proteins that probably interact with natural endogenous ligands. Much quoted examples are the opiate and ouabain receptors. The latter is the Na^+,K^+-ATPase, and active research is going on the purification of endogenous ligands that inhibit the Na^+ pump and regulate Na^+ fluxes in the kidney. In the case of the opiate receptors, the natural ligands (endorphins) are known. Studies are being conducted on the purification and mechanism of action of the opiate receptors, and functional reconstitutions should be achieved in the near future.

A. Drug Receptors

The ouabain receptor (Na^+,K^+-ATPase) was discussed in Lecture 6. An opiate receptor was first solubilized several years ago (Simmonds *et al.*, 1980; Rüegg *et al.*, 1980), and subsequent attempts of purification have been reported (Bidlack *et al.*, 1981; Cho *et al.*, 1983), but the preparations appear to be quite heterogenous. Reconstitution on addition of acidic phospholipids to a partially purified receptor resulted in a weak ligand binding activity suggestive of a μ-type opiate receptor (Cho *et al.*, 1983). It is surprising that the proteins solubilized by Triton X-100 were not purified in the presence of phospholipids, which are known to protect neuroreceptors as well as other receptors against inactivation. The availability of monoclonal antibodies against the opiate receptors should allow for more specific purifications in the near future. Several clues of interaction between opiate receptors and the adenylate cyclase system (cf. Gwynn and Costa, 1982) call for co-reconstitution experiments.

There are numerous other drug receptors that need to be explored by reconstitutions after suitable purification. I have mentioned earlier the purification of the slow Ca^{2+} channel and its interaction with a large number of antagonists and agonists which have become of considerable importance in the treatment of heart diseases. Two types of benzodiazepine receptors have been solubilized by differential extraction with detergents. The first one was extracted from membranes of brains of various animals with buffer containing 2% Triton X-100. The second was extracted from the insoluble residue by the same detergent in the presence of 1 M NaCl (Lo *et al.*, 1982). The solubilized receptors were immobilized and assayed by a novel lectin-agarose immobilization method (Nexo *et al.*, 1979). Amiloride-sensitive epithelial Na^+ channels were reconstituted into planar lipid bilayers by the osmotic gradient–facilitated fusion method (Sariban-Sohraby *et al.*, 1984). This assay may now serve for the purification of this important channel protein. The identification of the phorbol ester receptor with protein kinase C (Nishizuka, 1984) and of the cromolyn receptor with a Ca^{2+} channel, described earlier, are examples of using drugs to search for and purify receptors. How many more drugs like these will be used in the future? Cimetidine and chlorpheniramine are potential tools in the study of histamine receptors. What are the relationships between the antiepileptic valpoic acid to the GABA receptor, between chlorpromazine and related drugs and the dopamine receptors? How does tolbutamide stimulate the secretion of insulin? These and many other drugs used clinically (see *Physicians Desk Reference*) present us with challenges of a better understanding of drug use and a way to study new and old receptors. Many of them will be reconstituted, their natural ligands identified, and mode of action explored, in isolation and in co-reconstituted systems with appropriate effector proteins.

B. Toxin Receptors

The same kind of approach has been and will be used with numerous toxins that have been particularly useful in the study of neuroreceptors. Examples are α-bungarotoxin and scorpion toxin in the case of ACh receptor, tetrodotoxin and many other toxins mentioned above for the Na^+ channel, and apamin for the Ca^{2+}-dependent K^+ channel (Hugues *et al.*, 1982).

An interesting story is developing with the diphtheria toxin. We know a great deal about what happens once the A moiety of the toxin gets into the cell. We also know that the toxin can form channels in artificial lipid bilayers (Pappenheimer, 1977). There is only little known about the membrane receptor which interacts with the toxin. It appears that the B moiety of the toxin binds to the membrane and that Cl^- (or some other anions) are required for the entry of the toxin. SITS, an inhibitor of Cl^- transport, was shown to protect against diphtheria toxin (Sandvig and Olsnes, 1984). The toxin has been incorporated into planar lipid bilayers (Donovan *et al.*, 1981), forming voltage-dependent anion-selective channels, but only at low pH (about 5.0). It therefore seems likely that this property of the toxin may be relevant to a subsequent step, e.g., when the toxin is in the acid environment of endosomes. Thus, the first transmembranous movement of the A moiety remains a mystery.

A more coherent and fascinating picture has emerged from reconstitution studies with tetanus toxin (Borochov-Neori *et al.*, 1984). Following the old observations by van Heyningen (1963) that tetanus toxin interacts with gangliosides, these investigators demonstrated that the toxin forms channels in planar bilayers of phospholipids only when added to the compartment where the ganglioside is present. Of particular interest is the observation that, under these conditions, ion channels are formed by tetanus toxin at neutral pH. This is in contrast to diphtheria toxin and to tetanus toxin fragment B, which in the absence of gangliosides forms channels only at an acid pH (Boqúet and Dúflot, 1982). It seems quite likely that the formation of channels at neutral pH is closely related to the toxicity of tetanus toxin *in vivo* (M. Montal, personal communication).

Colicins and bacteriocins bind to receptors in the outer membrane of *E. coli*. Excellent reviews have been written on these interesting toxins (cf. Cramer *et al.*, 1983). Of particular interest for the reconstitutionists are the fundamentally different results depending on whether colicin K was incorporated into planar membranes or into liposomes (Luria and Suit, 1982). In liposomes, colicin K induced efflux of a variety of ions independent of a membrane potential, whereas in planar membranes ion flux depended on a potential (positive on the side of colicin addition). In view of the fact that none of these experiments was done in the presence of outer membrane receptors, the significance of these reconstitution experiments needs to be established.

Lecture 11

Reconstitutions of Pathological States

All roads lead to DNA.

We want to reconstitute pathological states for two reasons: to gain a better understanding of diseases, and to find methods aiding in their diagnosis and treatment. Reconstitution can take place at different levels. We can isolate genes responsible for a disease and sometime reconstitute the pathological state by transfecting a normal cell. The normal and transfected cells represent an ideal couple for comparative metabolic studies. In case of multigenic diseases, such as cancer, the situation is more complex, but as recent developments with oncogenes have shown, the same approach is feasible (cf. Cooper, 1982; Land et al., 1983; Weinberg, 1984). Time will overcome objections of artificiality raised by scientific naturalists. Abstract expressionism has arrived in biology, and if investigators fail to recognize the potential excitement of mixing genes from E. coli or yeast with those of mammalian cells, it will be their loss. Parallel to genetic transformations, we can reconstitute abnormalities phenotypically, an approach with a short past and a great future. Metabolic diseases, e.g., involving hormones or malnutrition, can be explored at the cellular level under complete control of environmental conditions. How do cells in tissue culture respond under growing or resting conditions to altered levels of insulin or prostaglandin? The work of Sato and his collaborators (cf. Barnes and Sato, 1980) on the differences in growth requirements of different cell types has opened up new avenues of approaches to the individuality of cells. Experiments of this type (Boerner and Saier, 1982) attain a level of precision that cannot be achieved with animals or perfused organs. Isolation and growth of cells, e.g., from different sections of the excretion machinery of the kidney, will allow us to examine the contribution of each player in this orchestra of functions. This approach may even lead to a revival of studies of bacterial infections at the cellular level. Parasites, protozoa, bacteria, fungi, and viruses that live inside cells (politely called their hosts) can be reexamined in such reconstituted systems at a new level of intimacy and specificity.

There is hardly a disease that does not involve, directly or indirectly, the structure and function of cellular membranes. A cell that is sick but has an unimpaired plasma membrane has a chance to survive. A cell with a membrane that leaks ions, unless quickly repaired, is doomed to die. In whole organisms, plants, of animals, individual cells are expendable; some more than others. The processes of differentiation, regeneration, and wound healing involve membranes and their functions. Models to study these processes at the membranous and cellular level are being constructed, including the reconstitution of nuclei, which I shall mention in my next lecture.

This lecture will deal mainly with one pathology, the malignant transformation of cells, simply because my own efforts during the past 30 years have focused on this phenomenon. The same kind of thinking and experimental approach is applicable to many other disease processes (e.g., cystic fibrosis) and many investigators are moving in the same direction, nebulous as it may appear at this stage. More of this in Lecture 12.

Lesson 34: A new pathology

Normality is such a miracle of coordination and interplay of metabolic controls that we should not wonder why we are sometimes sick, but why we are well most of the time. One price we pay for this well-being is that, because of its complexity, we do not understand it. A study of the history of medicine reveals that we often learn more about normality by investigations of pathology than of normality. Analyses of different cases of *Xeroderma pigmentosa* has taught us that even sick DNA has probably more lives than a cat (assuming an average of nine). In addition to clinical material, pathology has been broadened today by artificially generated mutant cell lines, a method which is taking another leap with site-specific mutagenesis. We can reconstitute pathology at all levels, phenotypically and genetically—in animals or plants, in cells, and in cell-free systems.

I. About the Artificiality of Cancer Research

A great deal has been written about cancer, and it is fashionable to emphasize its complexity and its staggering variability. The chemotherapeutic successes with specific types of tumors and drugs have reemphasized the fact that cancer is not a single disease and that hopes for a wonder drug to cure all forms of cancer are unrealistic. Nor is there a wonder drug against all infectious diseases. The similarity between cancer and infectious diseases in this and other respects is quite striking. They have in common the multitude of causative agents, a multitude of clinical manifestations, and an often underestimated variability in host susceptibility, which is influenced by a large number of nutritional and environmental components. We know a great deal about causative agents of infections; we have isolated most of the responsible agents and have "reconstituted" model diseases by infecting susceptible animals. The selectivity of these infectious agents for specific animals, specific organs, and specific cells is as astounding as the specificity of carcinogens and tumor promoters. As mentioned before, there is much more to learn about these specificities by studies at the cellular level. The similarity between cancer and infectious diseases becomes more apparent when we turn from man to animals that appear to be much more susceptible than man to malignancies by infectious viruses. In man, viruses causing cancer are much less frequent than mutated genes generated by environmental toxins and radiations. Over 40 years ago Stanley and Knight (1941) wrote a visionary article about the relationship of viruses and genes, which few took seriously. Today this relationship is taken for granted.

Our ignorance about the pathophysiology of cancer is actually not that much greater than our ignorance about the pathophysiology of infections. How much do we know about the biochemical effects of pneumococci on the lung or of hepatitis virus on the liver? Why do mice carry an encephalitis virus in their intestine, and only one mouse in many thousands catches the disease? Should we wait for such an event to study the disease? Only by isolating the virus from the stool and injecting it into the brain of a mouse (even of the same mouse) can we elicit the disease without fail, thus providing us with information of the process as well as with an assay for the purification of the infectious agent. What a wealth of information has emerged from such artificial studies of bacteria on agar plates and their metabolism in defined media and of their enzymes!

This discussion of infectious agents has a purpose. We admit that artificial transplantation and all *in vitro* or "*in plastico*" studies with growing cells are artifacts by definition. Yet, we emphasize their enormous value. We must reject objections that are being raised over and over again to the artificiality of modern cancer research, reject voices raised against the significance of oncogenes, against tumors generated by transplantations, against tumors generated in plastic dishes. We must reject them, but not discount them. We do know that 3T3 cells are not normal (Littlefield, 1982) and are not even human. But they are immortal cousins of mouse fibroblasts, perhaps only once or twice removed. Neither I nor the *New York Times* can overemphasize the importance of oncogenes isolated from human cancer by transfection into 3T3 cells (Weinberg, 1984). In addition to the wealth of information emerging from studies of their molecular biology, they will spare the lives and cost of thousands of mice and other animals. If these experiments inform us only about the last chapters in the cancer story, so be it. We can look forward to the excitement of earlier chapters of carcinogenesis with more virgin cells, possibly even of human origins. The advances will come. As will be discussed later, it is often easier in research to walk backward than forward.

II. Two Approaches to Cancer Research

If it is true that there is more than one way to skin a cat (though I fail to see the point of it) there are many more ways to skin a cancer cell. There are, however, two fundamentally different approaches. One I call the "DNA-to-XYZ route," the other the "ABC-to-DNA route." The first one is the quicker route and is currently very popular. It has allowed "reconstitution" of malignancy with multiple oncogenes and has yielded clues to their biological expression (Land *et al.*, 1983). It gave us also a great deal of information about the structure, location, and function of oncogene products. It has opened a new avenue to the study of the relationship between growth and protein tyrosine kinases. But XYZ, the biological substrates of the protein tyrosine kinase, the culprits presumably responsible for transformation, have remained elusive. Several suspects with short half lives have appeared on the scene but had to be released without bail. There are other difficulties with the DNA-to-XYZ approach. Of over 20 oncogene products (Bishop, 1983; Weinberg, 1984; Hunter, 1984), only about one-third catalyze tyrosine-specific phosphorylations of proteins. Three catalyze a threonine-specific autophosphorylation, but no other protein substrate has as yet been identified. Even an inactivated gene product (p21) did not serve as substrate (Riegler and Racker, unpublished observations, 1984). Some oncogene products have considerable homology with the src gene product, but no phosphorylated substrate has as yet been detected. One oncogene has a striking homology with PDGF; another with the EGF receptor. In some cases an interaction of the gene product with the plasma membrane is associated with transformation.

The identification of oncogenes and the possible relationship of oncogene products to protein kinases are among the most exciting discoveries in oncology during the past decade. How can we proceed with these discoveries? How can we identify the phosphorylated products that initiate the process of transformation, e.g., in Rous sarcoma virus–transformed cells? One of the problems with the src protein kinase is its lack of specificity. Many proteins are phosphorylated *in vivo* at a tyrosine residue in the transformed cell, including some glycolytic enzymes (Hunter, 1984). How can we tell whether the modification of any of these substrates is important? My distress about following this route of inquiry was enhanced by the finding (Braun *et al.*, 1984) that synthetic random polymers containing tyrosine and glutamic acid were much better (and cheaper) substrates for all protein tyrosine kinases tested, including the src gene product, than synthetic peptides custom-made according to the sequence analysis of the phosphopeptide in the src gene product. Thus, it seems to me that the task ahead on this road is to find a needle in a haystack without a magnet. It is more likely that luck rather than logic will succeed.

The second route illustrated in Fig. 11-1 is to choose what I call the scenic route from ABC to DNA. The altered DNA may or may

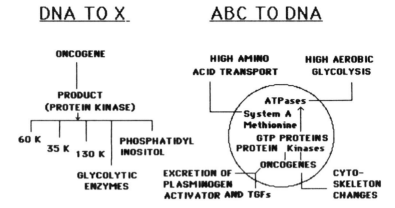

Fig. 11-1 Two biochemical approaches to diseases.

not be related to an oncogene (Sager, 1982). ABC represents biological or biochemical features characteristic of many, preferably most, transformed cells. By investigating any of these features in depth, we should eventually trace its origin to an altered gene. They may not lead us to a protein kinase or even to an oncogene—there may be converging routes—but they all should lead to an altered DNA. This is the route we entered some 30 years ago when we embarked on the Warburg phenomenon, the increased aerobic glycolysis of tumor cells. Before discussing this project and our attempts to reconstitute the Warburg effect, I want to talk about biological and biochemical properties of transformed cells that should be suitable as starting points on the road to DNA and to the design of reconstitution experiments.

I have listed in Table 11-1 some properties of tumor cells that differ from those of normal cells (cf. Land *et al.*, 1983). Immortality and loss of growth control represent what I believe to be the most important and compulsory features of transformed cells. A cell with

TABLE 11-1 **Biological Differences**

Normal Cell	Cancer Cell	Oncogene
Mortal (senescence)	Immortal (no senescence)	Adenovirus *E*1A, *Myc* Polyoma *Large T*, Abelson *v abl*
Control of cell cycle	Loss of control	Polyoma *Large T*, Rous *v src*, Abelson *v abl*
Not transformed	Transformed	*Ras*, Polyoma *middle T*, Rous *v src*, Abelson *v abl*

these two properties is a tumor cell, probably giving rise to a benign tumor. The other features I have listed I consider to be optional and related to the degree of malignancy, a viewpoint I shall elaborate on later. Recent advances in transfections have enabled us to make some tentative assignments of these features to specific oncogenes, as indicated in Table 11-1. That the transforming gene product of Rous sarcoma virus appears to induce at least two features, and that of the Abelson virus all three features, is of particular interest, in view of the fact that they catalyze phosphorylation of tyrosine residues in proteins and hydroxyl groups in other cellular components (Sugimoto et al., 1984; Macara et al., 1984). Before we accept, however, a too simplistic view, we must keep in mind that, e.g., the Rous sarcoma virus, in addition to generating its own messages, activates cellular genes (Groudine and Weintraub, 1980).

Hayflick and Moorhead (1961) have shown that primary human fibroblasts fade away (like old soldiers) after about 50 passages. Similar experiments were performed with primary mouse fibroblasts (Todaro and Green, 1963). These old soldiers (I mean the cells, not the authors) also faded away until a crisis took place and the immortal 3T3 cells emerged. (I am glad this does not happen to our generals.) The 3T3 cells are mutants: they are not normal, they do not age—they are invaluable.

What is aging? I was present at a lecture on this subject given by the great physiologist Walter B. Cannon in 1942. Frankly, the only thing I remember from his brilliant lecture is that when you grow old, hair starts growing from your nose and ears. Although I can confirm his observation, it does not help much. At a more biochemical level, we find in the literature descriptions of changes in membranes, in proteins, as well as in lipids, diminished functions of transport and appearance of new surface antigens (e.g., among senior citizens of erythrocytes). Or particular interest is the recent finding that senescent human fibroblasts have a diminished response to EGF, expressed by a loss of autophosphorylation of its receptor (Carlin et al., 1983). Kay et al. (1983) reported that a surface antigen of erythrocytes which appears in senescent cells before they are eliminated from circulation is a proteolytic fragment of the inorganic anion transporter, which has been shown to be phosphorylated (in vitro) at a tyrosine residue (Dekowsky et al., 1983). We have confirmed the in vitro (and not in vivo) phosphorylation observed by these investigators. In view of these observations, a possible connection between aging and protein tyrosine phosphorylation and dephosphorylation needs to be explored further.

An interesting fictional experiment was performed by a surgeon in a book by Lawrence Sanders called *The Sixth Commandment*. He extracted what he thought was an immortality factor from tumors with the hope to prolong life. In the story he transmitted tumors instead of immortality. But at a cellular level, similar experiments have been performed by transfection with immortality genes (E1-A or myc). We do not know how these genes accomplish this feature of immortality, and in fact they may do this indirectly. They may act by suppressing what I like to call a mortality gene. It is appealing to propose the existence of such a gene, a gene acquired during evolution, a gene important for evolution. Indeed it may be a gene related to a slow virus and I propose that it may act by influencing phosphorylation–dephosphorylation. This hypothesis calls for interesting experiments that are difficult, but feasible. Can we transfect a cancer cell with a mortality gene? Will it still be a cancer cell? Needless to say, I feel more comfortable with experiments on reconstitution of mortality than on immortality, and I am glad Sanders' surgeon failed.

The next biological feature of transformed cells is the loss of cell growth control. There are many laboratories working on this problem, isolating normal inhibitors and transforming growth factors that will be discussed later. It was proposed (Croy and Pardee, 1983) that there is a labile protein that regulates the cell cycle at the restriction point at G_1. In transformed cells, a labile protein of 68 kDa was shown to be stabilized and synthesized at a more rapid rate than in nontransformed cells. It would indeed be interesting to learn whether this putative control protein is regulated by phosphorylation. Another good candidate is a 53-kDa protein with a high turnover, proposed as a possible regulator of the cell cycle (cf. Shen *et al.*, 1983). A relation of these regulatory proteins to the myc gene product has not been demonstrated. It appears, however, that there are in some cell lines two sites of regulation at the G_0/G_1 phase, one at a primary step toward competence, the second during progression toward DNA synthesis (Pledger *et al.*, 1978). The former is influenced by some growth factors (e.g., PDGF), the second by others (EGF or insulin). C-*myc* expression has been shown to increase early after stimulation of quiescent cells, while c-*ras* Kirsten expression takes place in mid-to-late G_0/G_1 phase (Campisi *et al.*, 1984). These findings may be important clues to a better understanding of the relationship between specific oncogenes and growth factors.

What about the optional features observed in many tumor cells?

To find differences between normal and cancer cells is as easy as giving up smoking, according to Mark Twain. Oncologists have done it a thousand times. Yet most of their findings have little significance because they were anecdotal. Reported differences in enzyme activity or mobility of certain proteins in SDS–PAGE are difficult to evaluate. What is the proper control cell? How many tumors show the same characteristic differences? I have selected a few features that are common, though not obligatory nor unique characteristics of tumor cells. I shall discuss them briefly.

III. The Scenic Route from ABC to X

A. Reduced Dependency on Growth factors

When "normal" cells grow on a plastic surface they become confluent and stop growing. This phenomenon used to be called "contact inhibition" but was later renamed "density-dependent inhibition of growth" because it was found to be at least partly an artifact of convention. With 10% serum, normal cells stop growing when they become confluent, but with more serum they keep growing (Holley and Kiernan, 1968; Holley 1975). It was concluded that transformed cells, which continue to grow after reaching confluency, require lower levels of serum growth factors than normal cells.

This need for less growth factors could be caused by several mechanisms. Transformed cells (a) may indeed have a lower requirement; (b) their receptor may have a lower K_m for the ligand; (c) the lower requirement may not be real because transformed cells make their own growth factors; and (d) there may be loss of an inhibitor which in normal cells counteracts the effect of growth factors. Although all four mechanisms may be operative, I favor the last two because of available experimental evidence and their potential to elucidate the cancer problem (Temin et al., 1972; Burr and Rubin, 1975; De Larco and Todaro 1978).

It seems appropriate to point out here how little we know about growth factors. Some of them appear to have dual functions, called short-term and long-term effects. In the case of EGF and insulin, the rapid activation of the tyrosine-specific receptor PK results in autophosphorylation and phosphorylation of other intracellular proteins. Interestingly enough, some are phosphorylated at tyrosine, some at serine residues. Thus, a chain reaction of kinase activations appears to be operative. Some functional assays with growth factors also reveal a rapid response. Rozengurt and Heppel (1975) observed changes in K^+ uptake within minutes after addition of either serum, EGF or insulin to 3T3 cells. An increased Na^+ influx via the Na^+/H^+ antiporter, which I discussed in Lecture 9, appears to precede the increased uptake of K^+. The latter was therefore interpreted as being secondary to an increased intracellular concentration of Na^+, which activates the Na^+,K^+ pump (cf. Rozengurt 1983, 1984). But what are the connections between protein phosphorylation and the increased Na^+ influx? How does the higher intracellular pH resulting from Na^+ influx relate to the increased DNA synthesis? Interesting formulations have been offered (Moolenaar *et al.*, 1984), but some of their premises based on experiments with amiloride that were performed in the absence of bicarbonate have been challenged (Besterman *et al.*, 1984). Stimulation of glycolysis by EGF and other growth factors was observed in 3T3 and other cell lines (Diamond *et al.*, 1978; Racker *et al.*, 1984). These stimulations do not appear to be related to increased Na^+ pumping, since the accelerated glycolytic rates are not sensitive to ouabain. Obviously, even the early effects of growth factors cannot be explained by a common secondary effect. Everything in nature seems complicated until we search very hard and find that it is even more complicated than we first thought.

What about the late effects of growth factors? In the case of EGF, the early phosphorylation and the late mitogenic effects have been dissociated experimentally. A cyanogen bromide fragment of EGF (Schreiber *et al.*, 1981) or a monovalent FAB fragment of a monoclonal (IgM) antibody against EGF receptor, which competed with EGF, stimulated autophosphorylation of the receptor but had no mitogenic activity. Cross-linking of the cell-bound FAB fragment with anti-mouse antibodies resulted in receptor clustering, internalization, and mitogenic response. Another (IgG) monoclonal antibody, which did not compete with EGF, had no effect, but when cross-linked was fully active as a mitogen (Schreiber *et al.*, 1983). It appears therefore that the hormone site of the receptor does not have to be activated, that EGF itself is not required, that internalization is required to induce the effect in the nucleus, and that the early and late responses have different underlying mechanisms. However, it is conceivable that the separation of the early and late events is only apparent since quantitative differences in the rate of autophosphorylation were noted. Alternatively, the autophosphorylation of the membrane-bound EGF receptor may not be representative of the capacity of the kinase to phosphorylate other proteins in the cell. Perhaps travel of the endocytosed protein kinase is required for reaction with the nucleus. The *v-erb* B oncogene product, which has considerable homology with the EGF receptor as well as tyrosine PK activity (Gilmore *et al.*, 1985; Kris *et al.*, 1985), is an obvious candidate for such explorations.

In considering a role for an internalized growth factor in tumorigenesis, the oncogene (*v-sis*) of the Simian sarcoma virus is a suitable system. Its gene product, a glycosylated protein of 28 kDa, is a close relative of PDGF. In an excellent review, Hunter (1985) critically analyzes, with extensive documentation from the literature, the possible role of this oncogene product as an autocrine transforming growth factor. In some tumor cell lines the *v-sis* protein is excreted; in others it cannot be discovered in the medium, but requires a signal peptide to transform. An autocrine mechanism involving the interaction between the *v-sis* protein and the PDGF receptor appears therefore to be a valid hypothesis for future explorations. However, differences in the response of cells to PDGF and *v-sis* protein need to be explained.

A close relationship between the receptor for the growth factor CSF-1 and the c-fms oncogene (Sherr *et al.*, 1985) is yet another example for a possible link between growth factor receptors, oncogene products, and protein tyrosine kinase activity. All these findings are consistent with a sequence of events which begins with an interaction of specific growth factors with their receptors at the external side of the plasma membrane, followed by clustering and internalization that appear necessary for eliciting a mitogenic response. This formulation is far from complete and does not explain transformation as will be pointed out in the next lecture.

B. Increased Uptake of Nutrients

1. Glucose

There is ample documentation that many tumor cells take up glucose (or an analog of glucose) faster than normal cells. There is, however, limited information about the basic biochemical reason for this change. It appears that more glucose transporters are present on the surface of transformed cells (Salter *et al.*, 1982). We have shown (Fagan and Racker, 1978) that an increased glucose uptake can be induced by several means that have in common a greater cellular

ATPase activity. Exposure of cells for several hours to 2,4-dinitrophenol, which stimulates mitochondrial ATPase, or to arsenate, which stimulates glycolytic ATP hydrolysis, greatly increased the V_{max} of glucose uptake. We concluded that the increased glucose uptake merges with the trail of an increased glycolysis, both consequences of an increased ATP hydrolysis. How this is achieved is still unknown.

2. Amino Acids

Increased uptake of amino acid via system A is commonly seen in transformed cells (Foster and Pardee, 1969; Isselbacher, 1972). System A, which almost specifically transports nonmetabolizable MeAIB (Christensen et al., 1965) and less specifically alanine, methionine, and others, is under hormonal and nutritional control (Guidotti et al., 1978; Boerner and Saier, 1982). When cells are deprived of amino acids, system A becomes derepressed, i.e., a cycloheximide-sensitive increase in the transport activity is observed. Transformed cells have a very high transport activity even without amino acid deprivation and are usually not as markedly influenced by hormones. Normal MDCK cells require insulin for derepression, whereas prostaglandin promotes repression (Boerner and Saier, 1985). There are two interesting responses of tumor cells to amino acids that are transported by system A. MeAIB or methionine are more growth inhibitory to certain transformed cells than to their wild-type (better called mild-type) ancestors (Boerner and Racker, 1985), and I shall describe later experiments showing an inhibition of glycolysis of tumor cells by methionine. On the other hand, under some conditions tumors require considerably more methionine for growth than normal cells (Stern et al., 1984; Hoffman 1984). These observations are of obvious interest for the growth control and biochemistry of cancer cells.

C. Anchorage-Independent Growth and the Secretion of TGFs

Anchorage-independent growth is known to be one of the most characteristic features of transformed cells (Shin et al., 1975), although once again the correlation is not perfect (Gammon and Isselbacher, 1976; Marshall and Sager, 1981). Nevertheless, this feature has been useful for distinguishing tumor cells from normal cells, for the testing of drugs that inhibit growth of tumor cells in soft agar, and most importantly for the isolation of transforming growth factors. De Larco and Todaro (1978) discovered that tumors secrete polypeptides called transforming growth factors (TGF) that allow a line of normal kidney cells (NRK-49F) to grow in soft agar. The investigators at NIH have shown that two proteins are required, TGF-α and TGF-β (Roberts *et al.*, 1983). TGF-α is a 6,700-MW polypeptide with known sequence (Marquardt *et al.*, 1983). It is similar biologically to EGF, but has different structural and immunological properties. It binds to the EGF receptor and stimulates its autophosphorylation. Conversely, EGF replaces TGF-α in the soft agar assay. The interaction of TGF-α with the EGF receptor allowed for a rapid assay by competition with radioactive EGF. Whereas TGF-α is preferentially excreted by tumor cells, TGF-β is a constituent of normal cells and excreted into the medium by both normal and tumor cells. A comparison of the urine of normal persons and cancer patients revealed a characteristic increase in the cancer urine of some polymeric forms of TGF-α, but not of TGF-β (Kimball *et al.*, 1984). TGF-α and β are resistant to acid and heat, but sensitive to trypsin and exposure to dithiothreitol. In addition to stimulating growth in soft agar, TGFs imitate the transformed state with respect to changes in cytoskeleton (Ozanne *et al.*, 1980) and glycolysis (Racker *et al.*, 1984).

Sporn and Todaro (1980) have proposed that TGFs act as autocrine growth factors. Once excreted by a cell, they reattach at a specific site and induce transformation. This is an attractive hypothesis with respect to TGF-α. The mode of action of TGF-β appears more complex. Most investigators agree that much more TGF-α is excreted by tumor cells than by normal cells grown in tissue culture. Some transformed cell lines excrete much more TGF-β than their untransformed parent cell (Roberts *et al.*, 1984), but this does not seem to be the case in all cell lines (Tucker *et al.*, 1983; M. J. Newman and E. Racker, unpublished observation). As mentioned earlier, no differences in TGF-β excretion were observed in urines of normal persons and cancer patients (Kimball *et al.*, 1984). In spite of the observation that the acid-treated conditioned media of normal cells contain what appears to be TGF-β, these same cells do not grow in soft agar unless both TGF-α and TGF-β are added to the media. There are several alternative explanations for this apparent discrepancy and a large number of questions that need to be answered. Does the acid treatment of conditioned media release TGF-

β from a binding protein? Can a protease achieve the same? Is secretion of TGF-β different in anchored and nonanchored cells? Is there an alteration of the TGF-β receptor in transformed cells?

It is not clear how TGFs achieve the reconstitution of the phenotypically transformed state. Their biochemical link to the oncogenes is also unknown. A hypothesis suggesting a possible connection between phosphorylation and secretion of TGFs has been proposed (Racker, 1983b), and I shall return to this problem in my last lecture. The reconstitution of the transformed state with TGFs has opened up important avenues to both diagnosis and therapy of cancer.

D. Excretion of Plasminogen Activator

Like other optional features, plasminogen activator excretion is a frequent characteristic of tumor cells. Its study may not only lead us eventually to an altered DNA but should also yield new information on the process of protein secretion and its metabolic control. There are indications that plasminogen activator may not be excreted via secretory granules (O'Donnell-Tormey and Quigley, 1983). Twin experiments with normal cells before and after transfection with specific oncogenes or after phenotypic transformation with specific transforming growth factor could be most revealing.

These experiments gain in importance with the recent discovery that an antibody against one of the plasminogen activators inhibits the formation of metastases (Ossowski and Reich, 1983).

E. Changes in the Cytoskeleton

Dramatic changes take place in the cytoskeleton when cells are infected with, e.g., Rous sarcoma virus. Similar changes in the cytoskeleton can be provoked by injection of purified src protein tyrosine kinase (Maness and Levy, 1983). Although it was shown that vinculin becomes phosphorylated at a tyrosine residue (cf. Hunter, 1984), this takes place on such a small fraction of the total vinculin pool that the significance of these observations cannot be assessed at the present time. It seems likely that vinculin is yet another victim of the nonspecific nature of the src protein kinase. We therefore need to search for other phosphorylated members of the cytoskeleton. It is also possible that none of them will be altered, because the morphological transformation may be due to an effect mediated by phosphorylation of a soluble or plasma membrane component that influences cytoskeleton assembly.

I shall not dwell on some other optional changes that may take place on transformation. They all could serve as a starting point of a research trail towards an altered DNA. Particularly interesting are changes in surface glycolipids (Hakamori, 1981) and the observations that some gangliosides modulate both cell growth, binding of growth factors, and tyrosine phosphorylation (Bremer *et al.*, 1984).

F. The Warburg Effect

I have previously outlined (Racker, 1976) the history of this phenomenon (Warburg, 1930) which I shall not repeat here. I shall discuss, however, some more recent developments which bear on the significance of the high aerobic glycolysis of tumor cells. As in the case of other optional features, there are examples of tumor cells that do not exhibit a high rate of aerobic glycolysis. A glycolysis-less mutant of a tumorigenic cell line was shown to retain tumorigenicity when injected into nude mice (Pouysségur *et al.*, 1980). Although the authors conclude correctly that glycolysis is not essential for tumor formation, it should be pointed out that we need to be more quantitative when we speak about tumorigenicity. Indeed the data presented in this article suggest that the mutant cell line is less malignant than the parent cell. Dr. M. Newman in our laboratory observed that an established cell line of "normal" rat kidney cells gave rise to large tumors in nude mice several months after transplantation. But the incidence was low and the tumor grew slowly. Thus, the term *tumorigenicity* has little meaning without a quantitative evaluation. This has been lacking in earlier as well as in some recent studies, but was considered in others (Land *et al.*, 1983). The question is not whether the high aerobic glycolysis is a cause of cancer, which no one today is likely to suggest, but whether it influences the degree of malignancy, in view of the correlation between glycolysis and malignancy (Burk *et al.*, 1967). In any case, the very fact that an increase in aerobic glycolysis can be observed following transformation of a cell by a tumor virus is sufficient to qualify this phenomenon for the analysis by the ABC-to-DNA route. I have advocated this approach via the scenic route for several reasons. The most important one is that I believe that it will lead us successfully to the altered DNA that is responsible for this particular phenomenon. Indeed, it may lead to the discovery of an altered DNA or of an oncogene that thus far has escaped genetic analysis. A second reason is that we are likely to make, on the way to DNA, a number of observations that are of general biochemical interest. In the case of glycolysis, this approach involved us in the study of the basic mechanism of action and the efficiency of operation of the Na^+,K^+-ATPase and other ion pumps (Kuriki and Racker, 1976; Racker, 1976).

An increased rate of glycolysis alters the bioenergetic balance of the cell. A large proportion of ADP and P_i becomes diverted from the mitochondria to the glycolytic enzymes giving rise to a diminished respiration. We have called this the Crabtree effect (Wu and Racker, 1959) and pointed out that it is a counterpart to the Pasteur effect. The latter is expressed by an inhibition of glycolysis by mitochondrial oxidative phosphorylation, which dominates ADP and P_i utilization for ATP formation in most normal cells. A high aerobic glycolysis increases the flux of glycolytic intermediates and their excursions, via alternative routes, into biosynthetic pathways of nucleic acids, carbohydrates, lipids and proteins. Thus, some of the recorded changes in these macromolecules may be just consequences of an increased flux of glycolytic intermediates. Changes in the steady state of glycolytic intermediates may profoundly influence the physiology of the cell. To mention just one interesting example, the addition of pyruvate to the media of cells growing in tissue culture reversed the toxic effect of amiloride (Taub, 1978), an observation that needs confirmation and further exploration.

Our studies of glycolysis in a variety of tumor cell lines have led us to the following conclusions: An important rate-limiting factor in all tumor cells examined is the availability of inorganic phosphate. In most cells ADP is not limiting because hexokinase generates, on addition of glucose, sufficient ADP to sustain the initial rate of glycolysis. As pointed out previously (Racker, 1965, 1976), the intracellular P_i level in turn influences, by allosteric control, other rate-limiting enzymes such as phosphofructokinase and hexokinase. The intracellular ATP level is probably less influential as an allosteric effector, since it often does not change on transition from mitochondrial to glycolytic ATP generation. This fact is often ignored in discussions of tumor glycolysis and the Pasteur effect.

Another point of frequent misunderstanding is the role of ATPase in glycolysis. In any mammalian cell (irrespective of whether ATP hydrolysis is a rate-limiting step or not), it is compulsory that, for each lactate that is formed from glucose, one ADP and one P_i is converted to ATP. Thus *at steady state,* one ATP per lactate must be hydrolyzed to allow glycolysis to continue. Our investigations were focused on this problem: which of the many ATP-hydrolyzing processes in the cell contribute significantly to the P_i and ADP pool?

The contribution of mitochondrial ATPase was assessed by testing the inhibition of glycolysis by oligomycin or rutamycin. A small minority of cell lines showed inhibition of glycolysis by rutamycin (Racker 1976). It is possible that in these cells there is more of the 32-kDa uncoupler protein (Nicholls, 1983) or that a natural uncoupler, e.g., free fatty acid, is present in excessive amounts. Indeed, addition of sodium oleate or 2,4-dinitrophenol greatly increased rutamycin-sensitive glycolysis in every cell line we have tested. The analysis of inhibition by rutamycin is, however, complicated by the fact that, in many cell lines, rutamycin actually stimulates glycolysis by eliminating the mitochondrial competition for P_i and ADP (the Pasteur effect). It is therefore possible that loosely coupled mitochondria make a greater contribution to the P_i and ADP pool than we realize. In those resting cells that we have studied, biosynthesis of macromolecules did not seem to contribute much to this pool (Fagan and Racker, 1978).

A major contributor to the P_i and ADP pool in several (but not all) tumor cells is the Na^+,K^+-ATPase, as indicated by about 50% inhibition of glycolysis by ouabain (Racker, 1976; Racker et al., 1983). Are these cells enriched in Na^+,K^+-ATPase protein or does the pump work inefficiently? For many years we have convinced ourselves (but not many other investigators) that EAT cells have a defective and inefficient Na^+,K^+ pump. In desperation we even quoted Bob Dylan, "The pump don't work 'cause the vandals took the handles." The situation is not quite as bad, because in EAT cells the pump still works, but the vandals (oncogenes) only ruined the handles; the pump leaks. The evidence for this statement is as follows: First, we have calculated the pump efficiency by a method devised for erythrocytes (Whittam and Ager, 1965). The principle of this method is based on the established stoichiometry between lactate formed and ATP hydrolyzed. By measuring ouabain-sensitive lactate formation and $^{86}Rb^+$ uptake, an efficiency ratio of Rb^+/ATP of close to 2 was calculated. In erythrocytes, which have no mitochondria, this procedure was thus validated. In EAT cells the situation is more complex, since mitochondria compete for the P_i and ADP delivered by the Na^+,K^+ pump, resulting in an overestimation of efficiency. Elimination of mitochondrial competition and activation of the mitochondrial ATPase by dinitrophenol greatly diminishes the relative contribution by the Na^+,K^+ pump and usually makes the evaluation more complex and results in an underestimation of the efficiency ratio (Racker, 1976). Using the Whittam–Ager method, we calculated for EAT cells an efficiency of about 40%,

(Racker, 1981c). In the presence of quercetin, a bioflavonoid which inhibits the hydrolysis of the E_2—P intermediate of the Na^+,K^+ pump without interfering with the formation of $E_1{\sim}P$ (Kuriki and Racker, 1976), the efficiency ratio was greatly increased. The interpretation of these findings was presented elsewhere (Racker, 1984). Very recently we have extended these studies by measuring in EAT cells Na^+ efflux rather than Rb^+ influx (Racker and Riegler, 1985). These experiments were performed to eliminate objections which were raised that a Na^+ leak may account for our previous results with Rb^+. Once again we observed a low efficiency ratio in EAT cells that was greatly increased by quercetin. Of particular interest is the observation that low concentrations of neutral detergents rendered the Na^+,K^+ pump even more inefficient. In 3T3 cells, which do not exhibit ouabain-sensitive glycolysis, the addition of low concentration of NP-40 greatly increased glycolysis, which then also became partly sensitive to ouabain. Measurements of the effect of quercetin on Na^+ efflux in EAT cells revealed no significant change. Thus a low Na^+/ATP ratio in EAT cells could not have been caused by a Na^+ leak, which obviously would have resulted in an apparent high rather than low efficiency ratio.

The next experimental task was to reconstitute and elucidate the reason for the defective Na^+,K^+ pump. The report of a successful reconstitution of the defective Na^+,K^+ pump, rendered inefficient by phosphorylation of the ATPase by a cascade of protein tyrosine kinases (Spector et al., 1980), had to be withdrawn (Racker, 1981b, 1983d). There remained, however, two experimental observations that were pursued. The first one was the presence of a protein tyrosine kinase that phosphorylated the Na^+,K^+-ATPase of EAT cells. The kinase has now been partially purified (S. Nakamura, S. Braun, and E. Racker, unpublished experiments), but thus far no evidence has been obtained for a change in the biological activity of the pump enzyme; nor is there any evidence for any relationship between this enzyme and an oncogene product. Moreover, the presence of protein tyrosine kinases in growing or nongrowing and even nonnucleated cells (Tuy et al., 1983) has removed from this class of enzymes the distinction of being necessarily related to growth processes.

The second observation that we have pursued is the presence, in crude preparations of TGF, of an activator of a protein kinase in plasma membranes. We have confimed that these crude polypeptide preparations (a) stimulate NRK cells to grow in soft agar, (b) activate a protein kinase (PPdPK) that is present in plasma membranes of EAT and other cells (Racker, 1983a), and (c) stimulate glycolysis of several normal cells (Racker *et al.*, 1984). We have recently separated the heat-stable and trypsin-labile polypeptide that activates PPdPK from TGF activity and identified it as histone 1 (Abdel-Ghany *et al.*, 1984). Histone 1 did not stimulate glycolysis or anchorage-independent growth of NRK cells, but purified TGF-β preparations did. I shall therefore focus on the action of the latter.

Before describing the stimulation of glycolysis by TGF-β, I would like to expand on a pertinent problem mentioned earlier in this lecture. In many tumor cells glycolysis was not inhibited by either ouabain or by rutamycin. For example, chick embryo fibroblasts catalyzed glycolysis that was inhibited 50% by ouabain. However, after infection by Rous sarcoma virus glycolysis was increased, but the enhanced rate was not sensitive to ouabain (Fagan and Racker, 1978). Glycolysis in all tumor cell lines that we have tested was, however, sensitive to quercetin. This bioflavonoid is not a specific inhibitor. It inhibits many enzymes, including hexokinase, ATPases, protein kinases, and transport processes. Yet, at the concentration at which it inhibited glycolysis in intact cells, it had no effect on glycolysis in cell-free extracts or even in permeabilized cells in the presence of P_i and AMP (McCoy 1976). These experiments suggest that quercetin inhibits the release of P_i from either ATP or from other organic phosphate components of the cell or alternatively impairs the entry of P_i from outside. We have mentioned earlier that quercetin inhibits the hydrolysis of E_2—P (Kuriki and Racker, 1976), and we proposed that in intact EAT cells it increases the efficiency of the Na^+ pump by protecting E_2—P against the illicit entry of water. But how does quercetin inhibit glycolysis in cells that are not driven by an inefficient Na^+ pump?

We have recently made observations that might provide a clue to this question. We have described a new intracellular ATPase that is quite resistant to mitochondrial ATPase inhibitors and to ouabain, but is highly sensitive to quercetin. It is the chloride-dependent H^+-ATPase of clathrin-coated vesicles (Xie *et al.*, 1983). Once again we are in a dilemma as to whether the depletion of ATP by this and other organellar proton pumps is a significant factor in the overall energy budget. The maintenance of a very acidic pH may be costly only if the coated vesicles, endosomes, lysosomes, and other organelles with an acidic pH become leaky to protons. Are these pumps leaky in tumors, or do the pumps become less efficient for other reasons? We do not have as yet specific inhibitors for these organellar pumps that could help to answer these questions.

Let me return to the mode of action of TGF-β. We have shown recently that a pure preparation of TGF-β from human platelets stimulated glycolysis of NRK-49F cells (Boerner *et al.*, 1985). In independent studies a stimulation of glucose uptake in 3T3 cells was observed (Inman and Colowick, 1985). In both cases incubation of the cells with the growth factor for several hours was required and the effect was prevented by the presence of cycloheximide. For unknown reasons, crude preparations of TGF-β from placenta stimulated glycolysis more rapidly than pure TGF-β (Boerner *et al.*, 1985; Racker *et al.*, 1984). TGF-β also stimulated the uptake of MeAIB via system A. The most interesting observation was that system A substrates (MeAIB or methionine), but not other amino acids, strongly inhibited glycolysis in the presence of TGF-β (Boerner *et al.*, 1985). Moreover, 11 different transformed cell lines that were tested showed methionine-sensitive glycolysis (50–80% inhibition), whereas several "normal" cell lines tested showed much less or no inhibition under the same conditions. When NRK-49F cells were exposed to methionine in the presence and absence of TGF-β, the control cell was unaffected but the TGF-stimulated glycolysis was depressed to a level that was usually even lower than in the cells without TGF-β. This observation suggests that TGF-β not only increased the total available P_i pool, but also shifted vendors: The relative contribution from the methionine-sensitive supply increased. The inhibitory effect of cycloheximide suggests that TGF-β stimulates the synthesis of either a hydrolase (or an activator of a hydrolase) that increases the P_i pool or of a P_i transporter (or an activator of a transporter) that facilitates the entry of P_i. Alternatively, it may stimulate an enzyme that removes an endogenous inhibitor (e.g., methionine). These and other possible alternatives require further exploration.

To explore these phenomena in more clearly defined systems we have recently initiated a study of metabolic alterations induced by specific oncogenes. Rat-1 fibroblasts transfected with either *myc* or *ras* (Land *et al.*, 1983) were chosen for measurements of glycolysis and amino acid transport. The most dramatic alteration was a four- to sixfold increase in glycolysis in the ras cells while in myc cells glycolysis was similar to that in the parent rat-1 cells. On the other hand, myc cells exhibited an enhanced sensitivity to TGF-β that allowed us to develop a rapid assay for this growth factor by measuring the transport of radioactive MeAIB (Racker *et al.*, 1985). Ras cells were hardly affected by TGF-β. The observation of different metabolic effects induced by specific oncogenes may shed light on their cooperative effects on tumorigenicity (Land *et al.*, 1983).

Glycolysis in ras cells, but not of myc or rat-1 cells, grown to confluency was inhibited 50–70% by methionine. This inhibition required (a) a high concentration (10mM) of the amino acid, (b) incubation with methionine for several hours, and (c) the presence of serum during exposure to methionine. We observed that at high concentrations (10mM) methionine entered the cell by an order of magnitude faster than at 200 μM. Even at the high concentration, several hours were required before maximal intracellular levels of free methionine were reached. These findings are consistent with our proposal that a high intracellular concentration of methionine is required to inhibit glycolysis maximally. The serum which was required for the expression of methionine sensitivy could be substituted in the case of confluent ras cells by insulin, IGF II or most efficiently by IGF I. No other growth factor tested, including EGF, PDGF, and transferrin, was effective in substituting for insulin.

Our current hypothesis based on these observations is shown in Fig. 11-2. Methionine (or other substrates of system A) inhibits an ATPase that has become associated with the transport protein of system A, thus serving as an auxilliary motor for the import of methionine similar to the "moped" of the Na$^+$/H$^+$ antiporter of *S. faecalis* (Hefner *et al.*, 1980). The role of insulin would be to facilitate the movement of a masked ATPase from a "microsomal" location to the plasma membrane, analogous to its effect on the glucose transporter (Kono *et al.*, 1982). Of particular interest is how the ras gene product, a GTP-binding protein with a molecular weight of 21,000 (Scolnick *et al.*, 1979), induces a methionine-sensitive amplification of glycolysis. We propose, as depicted in Fig. 11-2, that it serves as a G protein between the transport protein of system A (representing the R protein) and the ATPase (representing the C

Hypothesis

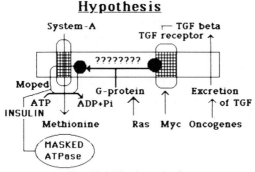

Fig. 11-2 The hypothesis.

protein), in analogy to the RGC systems discussed in Lecture 10. Similarly, we propose that TGF-β also sets in motion a mechanism whereby either an existing G protein or a newly synthesized G protein is activated to join the RGC system of the system A transporter. These speculations are subject to experimental verification or rejection. The seductiveness of this proposal is somewhat marred by several unanswered questions. Why is methionine much more effective as an inhibitor of glycolysis than other substrates of system A transport (Boerner and Racker, 1985)? Why do we see only a small increase of MeAIB transport in ras cells, which is so pronounced in other tumor cell lines? Does p21 represent a G protein that, because of its small size, fits improperly as a communicator yet can serve to activate the ATPase? Or do we need both myc and ras gene products to give rise to a high rate of amino acid transport? These questions are now being explored experimentally.

Finally, a comment on the relationship between oncogenes and TGFs. Why are these growth factors secreted in transformed cells? Is there an overproduction of the peptides as seems to be the case with TGF α, or do oncogenes give off signals (e.g., by phosphorylation) for the secretion of specific polypeptides as proposed earlier (Racker, 1983 a,b).

We have now come to a crossroad. In Ithaca, where I live, I have to go home via a traffic hazard called the "octopus." This is its common name, not only because of many car accidents, but also because there are eight different roads which funnel into it. Only two of them get me home quickly. A third one I could use, but it would probably take me three times as long. Most of the roads go away from home. The trouble with research is that there are many octopuses and we have only a vague idea where "home" is. Fortunately, that is the marvelous beauty about research. Although the wrong road may not lead us "home" (or whatever we call our primary research goal), it may lead you to places that are not as difficult to reach as "home" and worthwhile settling into. It is the mark of the gifted investigator to know what to pursue and what to ignore. The prayer of a young investigator could be a variant of the old one of F. C. Oetinger (1702–1782): "God, give me the serenity to drop a project I cannot do, the courage to change to a new one I can do, and the wisdom to know the difference."

Lecture 12

Glimpses into the Future of Reconstitutions: Hypotheses, Speculations, and Fantasies

> Scientists have odious manners,
> except when you prop up their
> theory; then you can borrow
> money from them.
>
> **Mark Twain**

You know how it goes: I formulate a hypothesis; you are speculating; he is fantasizing. In this lecture I have two collaborators, you and him. Since our styles of writing are quite different, there should be no difficulty in recognizing who is saying what. My hypotheses are always based on solid facts and suggest a number of exciting experiments that will be valuable for our understanding of the phenomenon. My collaborators do not always stick to the facts; they ignore some observations that simply do not fit their hypotheses. Their ideas are often based on intuition, which no one can nowadays take very seriously (though it may have been acceptable in the times of Newton, Pasteur, and Einstein). Yet, I will stand here to defend my two collaborators. They are quite imaginative, and once in a while their ideas are not too far off the mark. Do not laugh at them, just scrutinize what they say and decide for yourself whether their proposed experiments are worthwhile.

I. Methods of Reconstitution

I have discussed earlier the remarkable selective action of detergents on different membrane proteins. The ever-expanding syntheses of new detergents should greatly enhance our repertoire of specific applications. The apparently uniquely benevolent interaction between digitonin and the β-adrenergic receptor, between octylglucoside and the lactose and carnitine transporters, the superiority of cholate for reconstitution of cytochrome oxidase, and of Mega 9 or octylglucoside for bacteriorhodopsin mentioned earlier are just examples of what is likely to become a large list. Pedestrian advances should also come from greater attention in the application of various detergents to the conditions of pH, presence of monovalent or divalent cations, ionic strength, mode of detergent removal, etc. They can be as important in optimizing solubilization and reconstitution as in the crystallization of proteins. Syntheses of custom-made pure phospholipids with fatty acids of varying length and unsaturations should not only permit improved reconstitutions but also allow for new physicochemical studies of protein–lipid interactions, without contaminating natural products that often interfere with, e.g., fluorescence and tunable laser resonance Raman spectroscopy studies. There is an urgent need for new and rapid methods for the separation of reconstituted particles from the media, particularly when transport of nonionic or zwitterimic compounds are measured.

Ever improving patch and liposome patch-clamping methods (Hammill *et al.*, 1981; Tank *et al.*, 1982) should have a great impact on experiments designed to gain insight into the operation of single channels. Future developments in quantitation should allow assay and purification of elusive channels. Methods of driving ATP synthesis by an electric field (Rögner *et al.*, 1979; Gräber *et al.*, 1982) in reconstituted CF_1F_0 vesicles might prove useful in delineating the roles of Δ pH and Δ ψ.

Perhaps the most deficient methodology in reconstitutions, in spite of recent progress, is the preparation of giant functional proteoliposomes. Proteoliposomes (Racker *et al.*, 1979) or liposomes (Nazaki and Tanford, 1981) prepared by removal of octylglucoside have diameters of 2000 Å or greater and are more uniform than those obtained by freeze–thaw sonication. Yet they are still too small for insertion of electrodes and reconstitution of organelles. Giant liposomes have been prepared by a variety of methods (Deamer and

Bangham, 1976; Reeves and Dowben, 1969; Oku and MacDonald, 1983). In our hands none of these have thus far been suitable for reconstitutions of active transport systems, either because of inactivation of the proteins by the solvent or because of the fragility of the resulting proteoliposomes. Recently, Rigaud and his collaborators (Rigaud *et al.*, 1983, and Rigaud, personal communication) have been very successful in applying the reverse phase evaporation procedures of Szoka and Papahadjopoulos (1978) and of Darszon *et al.* (1980) for the incorporation of bacteriorhodopsin and CF_1F_0 from chloroplasts. The resulting vesicles are 0.2–0.7 μm in diameter and catalyze high rates of proton pumping. For the reconstitution of organelles and cells, a large internal volume is an essential feature. Among methods for preparation of such giant liposomes, the use of osmotic fusion in the presence of Ca^{2+} (Miller *et al.*, 1976), fusion with Sendai virus (Loyter and Volsky, 1979; Vainstein *et al.*, 1984; Gould-Fogerite and Mannino, 1985), and fusion with myristic acid between 17 and 20°C or lauric acid between 11 and 15°C (Kantor and Prestegard, 1975) are particularly promising. For studies with intermediate-sized proteoliposomes, the freeze–thaw procedure (without sonication) (Pick and Racker, 1979a) is quite suitable. It has been greatly improved and standardized by control of ionic and other conditions (Pick, 1981; MacDonald *et al.*, 1983). It allows for the preparation of liposomes and proteoliposomes of considerable size and stability and has been used for the reconstitution of the amino acid transporter (system A) from EAT cells (McCormick *et al.*, 1984) and of the lactate transporter from rabbit erythrocytes (M. J. Newman, S. Nisco, and E. Racker, unpublished experiments). It should be pointed out, however, that the expansion in size by water entry (MacDonald *et al.*, 1983) does not change the number of protein copies per vesicle. Methods for preparation of giant liposomes suitable for insertion of microelectrodes have been described by Rögner *et al.* (1979). The use of the incorporation procedure to insert proteins into already existing giant liposomes has also been used successfully (Zahlman *et al.*, 1980; J. L. Rigaud, personal communication). Some of these and other procedures have been discussed by Kagawa *et al.* (1982). The future is likely to bring innovations in the preparation of giant proteoliposomes that will yield stable and large vesicles that contain multiple copies and multiple species of protein resembling the composition of native membrane.

II. Orientation-Directed Reconstitution and Co-Reconstitutions

It will be important in the future to develop methods of directed protein insertion, e.g., for the reconstitution of mitochondria or chloroplasts. Formation of asymmetric proteoliposomes near the transition temperature of the phospholipids by cholate dilution (Wickner, 1976) needs to be extended to other proteins. The approach of random trials and errors has been partially successful in the reconstitution of a right side out atractyloside-sensitive ATPase complex (Banerjee and Racker, 1979) and of right side out bacteriorhodopsin vesicles (Happe *et al.,* 1977) but the data are far from satisfactory. There are several rational methods that should be developed in the future. One is to insert a hydrophobic leader polypeptide to the inner end of the inserted protein. Such a polypeptide could be a polyvaline or polyphenylalanine of appropriate lengths that would be covalently attached to insure preferential insertion, e.g., by the incorporation procedure. The second approach would be to attach a hydrophilic peptide to the outer end of the inserted protein, which will impair the inside out incorporation. Ideally such a molecule should be attached either by a reversible covalent modification or by selective antibodies with medium-low affinity, suitable for dissociation following reconstitution. As mentioned earlier, preliminary experiments with the photosystem I complex and a monoclonal antibody have been successful (N. Nelson and E. Racker, unpublished experiments, 1984). This field is wide open. Co-reconstitutions may turn out to be the most physiological way for site-directed reconstitution. We have seen striking differences in the incorporation procedure when a protein was inserted into liposomes prior to the incorporation of its next neighbor, e.g., in the respiratory chain (Eytan and Racker, 1977). Thus, directed reconstitution of only one member of a pathway in the right side out configuration may facilitate the proper orientation of the other members. More experiments need to be done with reversible cross-linkers that could be used for single-step, forced co-reconstitutions in the proper orientations. I think I shall give the credit for this idea to you. He is already planning to use isolated receptors (cloned in *E. coli,* of course) to be inserted into liposomes to guide, e.g., the precursor proteins of F_1 or cytochrome oxidase to a natural assembly. He is dreaming of all kinds of soluble factors that may be needed inside or outside the liposome to act as guides.

III. Mechanisms and Regulations

There is ample room for dreams in this section. Perhaps the most urgent new methods to be developed are the reversible dissociation of subunits from multisubunit complexes that have thus far met with failure. Among the most challenging complexes are the acetylcholine receptor and the respiratory complexes, including cytochrome oxidase. Even the reconstitution of bovine heart mitochondrial F_1 from its subunits has not been achieved by investigators who have successfully reconstituted F_1 from $E.$ $coli$ (cf. Lecture 5). Perhaps an analysis of the reasons for this failure would be most fruitful in view of the remarkable similarities between the two complexes. If we could learn to prevent the irresponsible hydrophobic interactions of mitochondrial F_1 or cytochrome oxidase subunits by increasing (preferentially reversibly) their hydrophilicity by covalent attachment of small charged compounds, we might gain information on how to separate subunits of enzymes that have not as yet been reconstituted from their subunits. A comparison of sequences and hydropathy plots of $E.$ $coli$ and yeast mitochondrial F_1 may guide us to a more rational approach to these problems. Also, more attention should be given to a role in reconstitutions played by cations and cofactors such as Mg^{2+} and ATP in F_1 and Cu^{2+} and heme in cytochrome oxidase. In the case of CF_1, reactivation of the enzyme dissociated in the cold showed a requirement for ATP (Lien and Racker, 1971). Stripteasing of individual subunits from CF_1 (cf. Richter et $al.,$ 1984) and from acetylcholine-receptor (Huganir et $al.,$ 1981) may turn out to be a slower but safer procedure for the exploration of the specific role of individual subunits, as it has been in the Salome dance of mitochondria, unveiling coupling factors. Affinity columns prepared with monoclonal antibodies against individual subunits or with specific interacting toxins may help in this approach. Eventually all these tedious approaches may be superseded by cloning and production of individual subunits, as in the case of acetylcholine receptor (Mishina et $al.,$ 1984; Fujita et $al.,$ 1985). Yet, the problems of irresponsible and nonspecific hydrophobic interactions need to be solved in any case prior to successful and efficient reconstitution. In all likelihood, both striptease resolution–reconstitution and assembly, starting with individual subunits, will yield the only definitive answers to the question of the role of individual subunits in the function of the complex. Yet let us be prepared for misleading answers as well. The reconstitution of an efficient Ca^{2+}

pump with a homogeneous 100-kDa ATPase from SR has been achieved with synthetic phospholipids (Knowles *et al.*, 1980; Navarro *et al.*, 1984). However, the lipid composition needed to achieve a high efficiency was very different from that present in the native membrane. When the optimal PE/PC ratio was substituted by a mixture of phospholipids resembling that of native vesicles, which contain an excess of PC, the pump functioned very poorly. Addition of a fraction from SR, enriched in proteolipids and phospholipids, to the PC vesicles during the reconstitution greatly enhanced the efficiency (E. Racker, E. A. Blair, and J. Navarro, unpublished observations).

This brings me to the problem of regulatory mechanisms. The regulation of ATP hydrolysis by the ϵ subunit of F_1 and CF_1 has been discussed previously (Racker, 1976). More sophisticated regulations take place between different transport systems. The regulation of utilization of lactose by enzyme III of the PTS system is one example (Saier, 1982). Further biochemical evidence for this allosteric mechanism by reconstitution experiments has been reported (Nelson *et al.*, 1983). A few regulatory phenomena between pathways, such as the Pasteur and Crabtree effects, were reconstituted many years ago (Gatt and Racker, 1959; Uyeda and Racker, 1965). There is a wealth of information to be gained by this approach about the interactions and regulations of crossing pathways between biosynthesis and degradation of carbohydrates, fatty acids, amino acids, and macromolecules. Questions such as the role of, e.g., citrate in the regulation of glycolysis need to be answered in reconstituted systems that closely imitate intracellular steady state conditions.

IV. Incorporations of Cellular Components into Cells

Similar to the fusion between proteoliposomes, the fusion of proteoliposomes with intact cells is favored by a high content in PE. Cytochrome oxidase proteoliposomes were incorporated into erythrocytes in the presence of Ca^{2+}; about 30,000–50,000 enzyme molecules were detected in each erythrocyte (Eytan *et al.*, 1979). Ca^{2+}-ATPase of SR was also incorporated into erythrocytes (about 20,000 molecules/erythrocyte) that were then capable of catalyzing an ATP-dependent uptake of Ca^{2+}. To eliminate the outward pumping of Ca^{2+} by the native erythrocyte pump, the cells were treated with

NEM prior to fusion. The enhanced rate of Ca^{2+} uptake is rather convincing evidence in favor of the proper (or rather improper) incorporation of the Ca^{2+} pump into the plasma membrane. These are important model experiments, but obviously not of physiological significance. One can dream up a large number of experiments that could be very informative, especially with respect to pathological states. Could we incorporate a defective Na^+,K^+-ATPase into normal cells and increase ouabain-sensitive glycolysis? Or better still, could we first inactivate the defective Na^+,K^+-ATPase in EAT cells and replace it with a tightly coupled pump, e.g., from dog kidney? A 175-kDa protein has been shown to be associated with the development of pleiotropic drug resistance of tumor cells (Ling *et al.*, 1983). Its function is unknown. Could this protein be reconstituted into liposomes and incorporated into a drug-sensitive cell in order to demonstrate its functional relevance? Such tests are feasible because drug-resistance appears to be closely correlated with a low-intracellular drug level following short-term exposure to radioactive colchicine or daunorubicin. Reconstitution of pathological LDL receptors (Brown *et al.*, 1983) could be useful in the analysis of its operations without secondary changes that they may have evoked during cell growth. Diseases of the plasma membrane with less clearly defined lesions could be approached by such reconstitution experiments. I have described in Lecture 10 how fruitful this approach has been in studies of RGC receptors.

The use of reconstituted Sendai virus envelopes as a vehicle for the microinjection of macromolecules and transfer of membrane component has vast possibilities (cf. Loyter and Volsky, 1979). For example, the inorganic anion channel of erythrocytes was incorporated into the plasma membrane of Friend erythroleukemia cells, which resulted in an increase of DIDS-sensitive $^{36}Cl^-$ flux. ^{125}I-Labeled cytochrome c injected by this procedure into Friend erythroleukemia cells distributed between the cell cytoplasm and the cell nucleus. Liposomes have been used to put a prokaryotic gene coding for β-lactamase into mammalian cells (Wong *et al.*, 1980). RNA (Wilson *et al.*, 1979) and DNA viruses (Fraley *et al.*, 1980) as well as immunoglobulins (Ostro, 1983) were incorporated into cells via liposomes. A rapid method for the preparation of highly fusogenic vesicles made of Sendai virus envelopes capable of delivery of DNA and IgG, as well as small-molecular compounds has been described (Vainstein *et al.*, 1984). The use of liposomes for drug delivery is outside the scope of these lectures on reconstitution, but in view of some of the misconceptions in this field, I would like to draw your attention to the level-headed review on this subject by George Poste (1983), which includes the exciting possibilities of targeting biological response modifiers to macrophages via liposomes.

V. Reconstitution of Organelles, Cells, Organs, etc.

Now we are approaching the inner sanctum of dreamland, but in some instances it does not seem that far away. Sarcoplasmic reticulum has two major capabilities: the storage and release of Ca^{2+}. The reconstitution of Ca^{2+} and P_i uptake, discussed earlier, has been achieved. The orientation problem has obligingly solved itself, since in the presence of excess phospholipids the vesicles are virtually all right side out. The major gap is the release of Ca^{2+}. Although a great deal is known about this process, the reconstitution of the responsible channel has still to be achieved. My invisible collaborators in this lecture tell me it will be done in the next few years. The next organelles in line for reconstitution are the clathrin-coated vesicles and endosomes. Their major task of providing an acidic interior has been successfully demonstrated in reconstituted vesicles (Xie *et al.*, 1984). The reconstitution of the Cl^- transporter is underway (D. K. Stone, personal communication). Purification and characterization of the DCCD-binding subunit of the proton pump needs to be done. There are other capabilities of these organelles that will be more difficult to elucidate. The most important one is the targeting of the endocytosed components (endocytosants?) to their future destinations. There are promising morphological studies suggesting modes of physical segregation (Willingham and Pastan, 1982), but at the biochemical level little is known. The trigger mechanism for endocytosis induced by ligand binding is also obscure. An interesting model has been described in a delightful letter (Haywood, 1983). Sendai virus was endocytosed by liposomes containing PC and gangliosides. The latter have previously been shown to serve as Sendai virus receptors. Binding of the Sendai virus took place in the cold, but "endocytosis" was observed only when the temperature was raised to 37°C. Other secondary functions of endosomes, such as iron delivery to the cytoplasm and to mitochondria, are simpler problems worthy of attack. But let us be guided by Piet Hein: "Problems worthy of attack prove their worth by hitting back." Another important aspect of clathrin-coated vesicles is the role of clathrin and its biological cycling. Decisive reconstitution experiments on the endocytotic process may help to solve some of the controversies on this subject.

Excellent progress is being made on the reconstitution of intracellular protein transport and posttranslational modifications during transit via the Golgi, as mentioned earlier. The near future should yield information not only on the details and mechanisms of these processes but also on the biological significance of the modifications such as glycosylation and phosphorylation that take place. Extensively purified Golgi fractions have been reconstituted and shown to require ATP and cytosol fractions for the transport of the vesicular stomatitis virus G-protein between compartments and for the resulting glycosylation (Balch *et al.*, 1984). With the tools of genetics and the advances in obtaining yeast mutants in the secretion process (Schekman *et al.*, 1983), we can also soon expect reconstitution experiments with complementation components. However, it seems likely that there exist multiple pathways of secretion and we should not neglect clues to alternative mechanisms that may not require packaging (O'Donnell-Tormey and Quigley, 1983). Secretory organelles, particularly those concerned with the delivery of biogenic amines and neurotransmitters, are high on my list of candidates for reconstitutions. The successful reconstitution of partial reactions, catalyzing transport of GABA or adrenalin mentioned earlier, are first steps in this task.

Once we have mastered oriented reconstitutions into giant liposomes and learned how to incorporate the receptors of protein import into phospholipid bilayers, together with the required proteases, we might begin to simulate a mitochondrion.

The exciting beginnings in the reconstitution of nuclei (Forbes *et al.*, 1983) are opening a new field of exploration. Let us also remember that Danielli, one of the "fathers" of membranology, ventured into the field of cell reconstitution (Jeon *et al.*, 1970). More of this and the reconstitution of organs in my next series of lectures (1994). But the success of using skin reconstituted in plastic dishes in the treatment of burns (Gallico *et al.*, 1984) suggests again that some dreams in this area have already come true.

VI. Reconstitution of Pathological States

A. Future of Studies with Cell Lines

Allow me and my two collaborators to plunge into this dreamland of unlimited possibilities. Cells are being isolated from patients who have diseases with the same name but with different genetic and metabolic lesions, and each of these cells may have their private requirements of optimal growth conditions. Obviously, a muscle cell from patients with Luft disease (Luft *et al.*, 1962; DiMauro *et al.*, 1976) with a defect in respiratory control of the mitochondria may be more difficult to grow than its normal sister cell. Once the goal is achieved, however, it will be possible to establish with how many disease processes we are dealing. If the mitochondria are "loosely coupled," we can explore specific possibilities. Is there an increased activity of the 32-kDa uncoupling protein (Nicholls, 1983), or are natural uncouplers of the fatty acid type allowed to accumulate, or is the membrane leaky because of improper phospholipid assembly? Or, is there an increased cycling of Ca^{2+}, as suggested by DiMauro *et al.* (1976)?

During the past decade there has been an exponential increase in publications describing enzymes and membranous changes during disease processes (Galjaard, 1980; Conzelmann and Sandhoff, 1984). A comprehensive review by Carafoli and Roman (1980) has covered mitochondrial changes associated with a large number of diseases. As appropriately pointed out by the authors, a primary involvement of mitochondria has been documented in only a minority of diseases. Though many of these diseases are rare, the future will bring a surge of efforts of probing into these diseases by biochemical and genetic means. The increasing number of cell lines that are representative of diseases will allow not only a depth analysis of the lesions but also the identification and isolation of the diseased genes with parallel efforts of curing the disease at the cellular level by reconstitution with the healthy genes. In the past, the major obstacles in the isolation of appropriate cell lines have been difficulties in the maintenance of differentiated normal and diseased cells in tissue culture. With recent advances in growing cells in defined medium (in the presence of cell-type specific growth factors and in the absence of cell-type specific inhibitors that are present in standard "fetal calf serum," favored by tissue culture experts of yesterday) the field has reached a stage—if not of adulthood—of promising adolescence. Use of monoclonal antibodies against undesirable cells will facilitate the isolation and growth of wanted cell types. Increasing numbers of diseased cell lines will be grown, if necessary, after immortalization by oncogenes. They will then be studied in peace without repeated biopsies and pain and danger to the patient. To aid in such studies, great advances are being made in the miniaturization of biochemical analyses of cultured cells (Galjaard, 1982). Nevertheless, I speak of adolescence of the field to avoid an impression of over-simplification. The clinical heterogeneity of diseases, the need for complementation analyses, the genetic heterogeneity, and the existence of different isozymes in different cell types (Galjaard and Reuser, 1984) are staggering complications.

An example of the difficulties associated with the assignment of a primary mitochondrial lesion to a disease process is the classical case of the Warburg hypothesis of damaged respiration as the origin of cancer. Few will dispute the experimental fact that in some tumors mitochondria may be damaged or may be fewer in number compared to normal cells. Such happenings contribute to an increase in the glycolytic rate. However, the very fact that in many tumor cells addition of rutamycin greatly enhances glycolysis shows that the mitochondria are very much alive and plenty in numbers, so that they effectively compete for P_i and ADP with the glycolytic pathway. Thus, observed changes in tumor mitochondria may be secondary to increased glycolysis rather than the other way around. My decision to choose glycolysis as a suitable start on the route to DNA was partly based on the fact that an increase in aerobic glycolysis is an almost universal feature following transformation.

The "reconstitutions" of transformations, both genetically by transfection and phenotypically by transforming growth factors, are experimental facts. Whether or not these model tissue culture systems are acceptable to biological naturalists is irrelevant, because they have already been fruitful at the level of diagnosis and treatment of cancer. The near future will bring a plethora of new drugs discovered by clonogenic (Salmon and von Hoff, 1981) or other rapid "in plastico" assays (Weisenthal et al., 1983). The future will bring a flood of biological response modifiers that will be analyzed as putative chemotherapeutic agents against specific cancers. Growth factors and their receptors will be cloned and mass-produced and chemically modified to serve as competitors. A host of natural antagonists of growth (including TGF-β) that are present in normal tissues and fluids will be isolated and tested in assays of growth in soft agar and in developing new biochemical approaches, which I shall discuss later. Most of these attempts will end up as failures in terms of clinical applicability, but a few will be successful in one or the other type of tumors. I hope we shall be able to devise more models for the formation of metastases, one of the key problems in oncology. We need to examine many problems in the simplest available model systems. What is the reason for the heterogeneity of tumor cells. (Fidler *et al.*, 1981; Poste, 1983)? How do muramyl peptides activate macrophages that kill tumor cells selectively? Why is a monoclonal antibody against plasminogen activator effective in combatting metastases in one particular model system (Ossowski and Reich, 1983)?

B. Reconstitution of Cancer: Hypotheses and Plans

Our own hypotheses on which we are basing future experiments are as follows: (1) We accept tentatively the autocrine formulation of Sporn and Todaro (1980) as a working hypothesis. TGFs are excreted by tumor cells and are responsible for many or all features of transformation. I say tentatively because TGFs are excreted by normal cells and found in normal human urine (Kimball *et al.*, 1984). (2) We accept the formulation of oncogenes as initiators of the events leading to transformation, but I shall return later to an alternative possibility. (3) We propose (Racker, 1983a,b) that in cancer cells there is an imbalance between phosphorylation and dephosphorylation reactions, both at the nuclear level (activation of genes) and regulations of the cell cycle and at the cytoplasmic and plasma membrane level (e.g., activation of an excretion process). Since normal human urine contains both TGFα and β (Kimball *et al.*, 1984), we need to postulate quantitative or qualitative differences among these growth factors with respect to polypeptide structure and aggregation, as well as in the induction of their excretion from the cell. We also have to consider changes in their plasma membrane receptors. In view of the fact that so many oncogene products have protein kinase activity (Bishop, 1983; Hunter, 1984), obvious speculations are that phosphorylation can directly or indirectly signal secretion of TGFs, change their structure or their form of aggregation, and may induce alterations in receptors. There is increasing evidence emerging for a link between receptors and oncogene products. The similarity between G-proteins of the plasma membrane and the ras gene product, pointed out to me in 1980 by Dr. Z. Selinger, is now a matter of great excitement.

We shall be searching for differences in the excretion of phosphorylated proteins. Such differences have already been recorded in the literature. A 35-kDa phosphorylated glycoprotein (Gottesman and Cabral, 1981) and a 60-kDa phosphorylated protein (Senger *et al.*, 1983) have been described as specific excretion products of transformed cells. Although experiences in our laboratory (M. Newman, unpublished) are not consistent with the statement that the 60-kDa phosphoprotein is preferentially secreted by transformed cells, the recorded observations need to be explored further. Of particular interest would be a possible association between these excreted phosphoproteins and a biological function of either stimulation or inhibition of growth. What about the intracellular events that follow transformation? What is responsible for the high-aerobic glycolysis and what is its relation to TGFs and oncogene products? What experiments can we design to reconstitute the Warburg effect? We believe that our best chances in the future are to pursue our findings on the effect of TGF-β on glycolysis and amino acid transport. We have observed a stimulation of glycolysis and of uptake of radioactive MeAIB by addition of TGF-β to normal cells (Boerner *et al.*, 1985; Racker *et al.*, 1985). Others (Inman and Colowick, 1985) have observed stimulation of glucose uptake by TGF-β. We have previously suggested that stimulations of both glycolysis and glucose uptake (Scholnick *et al.*, 1973; Fagen and Racker, 1978) are secondary to an increased generation of inorganic phosphate. How is this achieved? In the case of EAT cells and a few other tumors, we have shown that an inefficient Na^+, K^+ pump hydrolyzes excessive ATP (Racker, 1976, 1985). In the case of other tumors, an increased intracellular availability of P_i may be secondary to the increased activity of another ATPase or to a pyrophosphatase or to a protein phosphatase, or to a futile cycle, or to an increased uptake of P_i. We and others (Racker, 1965; Cunningham and Pardee, 1969) have presented some experimental evidence in support of an increased P_i uptake in tumor cells.

I have stressed here and previously the importance of P_i as a rate-limiting factor in glycolysis. Obviously ADP is equally important in the steady state, and alternative routes of its regeneration should be mentioned: hexokinase, phosphofructokinase, protein kinases, etc., all generate ADP. When hexokinase generates ADP, glucose 6 phosphate is formed, which inhibits hexokinase and therefore must be removed. This can be achieved by phosphofructokinase or by glucose-6-phosphatase. The latter is a member of a futile cycle that generates both P_i and ADP and increases glycolysis with no net energy gain. During cell growth, P_i is incorporated into macromolecules that turn over relatively slowly. If insufficient P_i is available and the synthesis of ATP becomes rate-limiting, the energy charge of the cell drops and the cell loses its viability. Cells must take out all kinds of insurances that this cannot happen; that their energy charge is maintained. Can we induce an immortal cell to be converted to mortality by transfection with a mortality gene, and can we induce a mortal cell to divide indefinitely by adding to the medium appropriate immortality growth factors? A mortal cell may fade away because it is starving of intracellular P_i. In contrast, an immortal cell will continue to grow and not fade away as long as it has enough P_i to feed on (among other things). This speculation suggests a new experimental approach to the problem of aging. The entry and exit of P_i into the cell is the first control station. The incorporation of P_i into intracellular components and the return of P_i by hydrolysis of phosphorylated macromolecules or storage phosphates is the second control station. A cell might age if the control stations are out of step. A cell can starve by limited P_i formation or import, or it can choke by excess formation of organic phosphate. These balances can be analyzed experimentally by measuring ATP turnover. I am not aware that they have been considered as factors in the process of aging. Primary mortal cells grown in tissue cultures and their transformed cousins are a good model system for such investigations.

We have, I believe, a new clue to the problem of high aerobic glycolysis in tumor cells. The observation that glycolysis in transformed cells or in untransformed cells that were exposed to TGF-β is sensitive to methionine allows us to design decisive experiments. A survey of the literature revealed some interesting observations on methionine which add up to what I call the methionine paradoxes. Compared to normal cells, many tumors have a low intracellular methionine level (Stern *et al.*, 1984), but an increased rate of methionine uptake via system A as mentioned in Lecture 11. Tumor cells have a higher nutritional requirement for methionine (Hoffman, 1984), but methionine and other substrates of system A are more growth inhibitory for transformed than for nontransformed cells

(Boerner and Racker 1985). Finally, methionine deficiency induces tumor formation (cf. Hoffman, 1984). It thus seems that many tumors have a narrow methionine window: too little is not enough, too much is toxic for growth. Although both features lend themselves to therapeutic approaches, it is obvious that rigorous control of methionine levels needs to be ascertained.

How can the methionine paradoxes be explained in biochemical terms? One possibility is that methionine is more rapidly utilized and/or excreted (e.g., as homocysteine). This would be consistent with the faster rate of uptake and the need for more methionine. The more rapid utilization could result in the accumulation of toxic metabolites, hence the greater sensitivity to high methionine concentrations. Enhanced transmethylation and increased excretion of homocysteine in tumors have been recorded (German *et al.*, 1983; cf. Hoffman, 1984).

How do these explanations relate to the observations on glycolysis? We propose that the intracellular level of methionine has a regulatory function on glycolysis. High concentrations in normal cells could suppress; low concentrations in tumors could enhance glycolysis. This simplistic explanation of the Warburg effect is at best only part of the story. We observed (unpublished experiments) that homocysteine counteracts the inhibition of glycolysis by methionine. Thus, the ratio of methionine to homocysteine and the ratio of their *S*-adenosyl derivatives (Hoffman, 1984) could be more critical than the concentration of methionine itself.

What is the mechanism of the bizzare inhibition of glycolysis by methionine? In view of the subtitle of this lecture, we take the liberty to propose that methionine inhibits an ATPase associated with system A of amino acid transport as outlined in Lecture 11. Although reconstitution experiments have shown Na^+-driven methionine uptake takes place in the absence of ATP, we can invoke the operation of a "moped", an auxiliary ATP-dependent accelerator, as described for the Na^+/H^+ antiporter in *S. faecalis* (Heefner *et al.*, 1980). This proposition is not unreasonable in view of the specificity of system A substrates as inhibitors of glycolysis. Moreover, this formulation suggests a number of experiments that may prove or disprove its validity. We are conducting experiments aimed at demonstrating an effect of methionine on ATP turnover, and we are designing experiments that may lead to the demonstration of a methionine-sensitive ATPase and an ATP-stimulated methionine transport. We are also designing experiments to identify the proteins that are synthesized at an enhanced rate during a few hours of exposure of normal cells to TGF-β. We are designing experiments to identify the proteins that interact with methionine to cause inhibition of glycolysis. We are performing experiments (described in Lecture 11) analyzing which oncogene induces a high rate of glycolysis and of MeAIB uptake. We are exploring the possibility that p21, the transforming product of the *ras* gene, participates as a G protein in system A transport of amino acids. We are looking for a link between TGF-β and a G protein. We are considering experiments that may shed light on the specific action of insulin which allows expression of the inhibition of glycolysis by methionine.

We are in the middle of experiments exploring the possibility that not every road from ABC to DNA will get us to an oncogene. I have suggested earlier (Racker, 1983b) that there is an attractive alternative. Oncogenes generate proteins that are either altered or produced in increased amounts. The proteins may be enzymes or part of receptors or may interact with the nucleus to activate genes. An alternative to an oncogene is an altered DNA that has lost its function or produces a defective product. There are numerous descriptions in the old and more recent literature of growth inhibitors that are present in normal tissues. Although in the past these inhibitors have escaped definitive isolation and characterization acceptable to the scientific establishment, they are bound to be caught in the very near future with the aid of modern isolation procedures. A rather cell-specific inhibitor was found to be excreted by monkey kidney cells. It has an M_r of 24 kDa, inhibits Na^+ influx (Walsh-Reitz *et al.*, 1984), and is now reported to be closely related to TGF-β (Tucker *et al.*, 1984). We are studying some inhibitors of growth in soft agar that we and others (Iwata *et al.*, 1983; Levine *et al.*, 1984) have found to be present as excretion products of tumor cells and that can be separated from TGF-β. We are finding, in placenta and in other normal cells, including some microorganisms, inhibitors of cell growth in soft agar. The possibility of cancer as a loss of a suppressor gene has some eloquent proponents (Marshall and Sager, 1981; R. Sager, personal communication).

A few years ago I proclaimed that should I succeed in the reconstitution of the Warburg effect *in vitro,* I would be ready to retire (Racker, 1983a). Now that we are coming a little closer to this goal, now that I realize that this will be just a beginning to a better understanding of cancer pathology and that it may give us new important clues to normality, I am inclined to withdraw my proclamation.

Bibliography

Abdel-Ghany, M., Riegler, C., and Racker, E. (1984). *Proc. Natl. Acad. Aci. U.S.A.* **81,** 7388–7391.

Abrams, A. (1965). *J. Biol. Chem.* **240,** 3675–3681.

Ahlers, J., Ahr, E., and Seyfarth, A. (1978). *Mol. Cell. Biochem.* **22,** 39–49.

Alfonzo, M., and Racker, E. (1979). *Can. J. Biochem.* **57,** 1351–1358.

Alfonzo, M., Kandrach, M. A., and Racker, E. (1981) *J. Bioenerg. Biomembr.* **13,** 375–391.

Aloj, S. M., Kohn, L. D., Lee, G., and Meldolesi, M. F. (1977) *Biochem. Biophys. Res. Commun.* **74,** 1053–1059.

Alper, S. L., Beam, K. G., and Greengard, P. (1980) *J. Biol. Chem.* **255,** 4864–4871.

Andersen, J. P., Skriver, E., Mahrous, T. S., and Møller, J. V. (1983). *Biochim. Biophys. Acta* **728,** 1–10.

Anderson, D. J., and Blobel, G. (1981). *Proc. Natl. Acad. Sci.* **78,** 5598–5602.

Anderson, W. M., and Fisher, R. R. (1978). *Arch. Biochem. Biophys.* **187,** 180–190.

Anderson, W. M., and Fisher, R. R. (1981). *Biochim. Biophys. Acta* **635,** 194–199.

Andreo, C. S., Patrie, W. J., and McCarty, R. E. (1982). *J. Biol. Chem* **257,** 9968–9975.

Anholt, R., Lindstrom, J., and Montal, M. (1980). *Eur. J. Biochem.* **109,** 481–487.

Anholt, R., Lindstrom, J., and Montal, M. (1981). *J. Biol. Chem.* **256,** 4377–4387.

Anholt, R., Lindstrom, J., and Montal, M. (1984). *In* "The Enzymes of Biological Membranes" (A. Martonosi, ed.), 2nd ed., Vol. 3, pp. 335–401. Plenum, New York.

Apps, D. K. (1982). *Fed. Proc, Fed. Am. Soc. Exp. Biol.* **41,** 2775–2780.

Apps, D. K., and Schatz, G. (1979). *Eur. J. Biochem.* **100,** 411–419.

Arion, W. J., and Racker, E. (1970). *J. Biol. Chem.* **245,** 5186–5194.

Ashani, Y., and Catravas, G. N. (1980). *Anal. Biochem.* **109,** 55–62.

Ashwell, G., and Harford, J. (1982). *Annu. Rev. Biochem.* **51,** 531–554.

Avron, M. (1963). "La photosynthèse," pp. 543–55. CNRS, Paris.

Azzi, A. (1980). *Biochim. Biophys. Acta* **594,** 231–252.

Azzi, A., Bisson, R., Casey, R. P., Gutweniger, H., Montecucco, C., and Thelen, M. (1979). *In* "Membrane Bioenergetics" (C. P. Lee, G. Schatz, and L. Ernster, eds.), pp. 13–20. Addison-Wesley, Reading, Massachusetts.

Baginsky, M. L., and Hatefi, Y. (1969). *J. Biol. Chem.* **244,** 5313–5319.

Baird, B. A., and Hammes, G. G. (1979). *Biochim. Biophys. Acta* **549,** 31–53,

Baker, P. F. (1976). *Fed. Proc., Fed. Am. Soc. Exp. Biol.* **35,** 2589–2595.

Bakker, E. P., and Harold, F. M. (1980). *J. Biol. Chem.* **255,** 433–440.

Balch, W. E., Dunphy, W. G., Braell, W. A., and Rothman, J. E. (1984). *Cell* **39**, 404–416.

Baldwin, S. A., and Lienhard, G. E. (1980). *Biochem. Biophys. Res. Commun.* **94**, 1401–1408.

Bamberg, E., Hegemann, P., and Oesterhelt, D. (1984). *Biochim. Biophys. Acta* **773**, 53–60.

Banerjee, R. K. (1981). *Mol. Cell. Biochem.* **37**, 91–99.

Banerjee, R. K., and Racker, E. (1979). *Membr. Biochem.* **2**, 203–225.

Banerjee, R. K., Epstein, M., Kandrach, M., Zimniak, P., and Racker, E. (1979). *Membr. Biochem.* **2**, 283–296.

Barchi, R. L. (1983). *J. Neurochem.* **40**, 1377–1385.

Barnes, D., and Sato, G. (1980). *Anal. Biochem.* **102**, 255–270.

Barsky, E. L., Dancshazy, Z., Drachev, L. A., Il'ina, M. D., Jasaitis, A. A., Kondrashin, A. A., Samuilov, V. D., and Skulachev, V. P. (1976). *J. Biol. Chem.* **251**, 7066–7071.

Barzilai, A., Spanier, R., and Rahamimoff, H. (1984). *Proc. Natl. Acad. Sci. U.S.A.* **81**, 6521–6525.

Baumann, H., Hou, E., and Doyle, D. (1980). *J. Biol. Chem.* **255**, 10001–10012.

Bäumert, H. G., Mainka, L., and Zimmer, G. (1981). *FEBS Lett.* **132**, 308–312.

Bayley, H., Huang, K.-S., Radhakrishnan, R., Ross, A. H., Takagaki, Y., and Khorana, H. G. (1981a). *Proc. Natl. Acad. Sci. U.S.A.* **78**, 2225–2229.

Bayley, H., Radhakrishnan, R., Huang, K.-S., and Khorana, H. G. (1981b). *J. Biol. Chem.* **256**, 3797–3801.

Beattie, D. S., and Villalobo, A. (1982). *J. Biol. Chem.* **257**, 14745–14752.

Beechey, R. B., Hubbard, S. A., Linnett, P. E., Mitchell, A. D., and Munn, E. A. (1975). *Biochem. J.* **148**, 533–537.

Belt, J. A. (1983). *Mol. Pharmacol.* **24**, 479–484.

Belt, J. A., Jarvis, S. M., Paterson, A. R. P., Tse, C. M., Wu, J. S., and Young, J. D. (1984). *J. Physiol. (London)* **353**, 87P.

Bengis, C., and Nelson, N. (1975). *J. Biol. Chem.* **250**, 2783–2788.

Bengis, C., and Nelson, N. (1977). *J. Biol. Chem.* **252**, 4564–4569.

Bennett, A. B., O'Neill, S. D., and Spanswick, R. M. (1984). *Plant Physiol.* **74**, 538–544.

Benz, R., Janko, K., Boos, W., and Läuger, P. (1978). *Biochim. Biophys. Acta* **511**, 305–319.

Berden, J. A., and Henneke, M. A. C. (1981). *FEBS Lett.* **126**, 211–214.

Berger, E. A., and Heppel, L. A. (1974). *J. Biol. Chem.* **249**, 7747–7755.

Berridge, M. J., and Irvine, R. F. (1984). *Nature (London)* **312**, 315–321.

Besterman, J. M., Tyrey, S. J., Cragoe, E. J., Jr., and Cuatrecasas, P. (1984). *Proc. Natl. Acad. Sci. U.S.A.* **81**, 6762–6766.

Bidlack, J. M., Abood, L. G., Osei-Gyimah, P., and Archer, S. (1981). *Proc. Natl. Acad. Sci. U.S.A.* **78**, 636–639.

Bidlack, J. M., Ambudkar, I. S., and Shamoo, A. E. (1982). *J. Biol. Chem.* **257**, 4501–4506.

Birchmeier, W., Kohler, C. E., and Schatz, G. (1976). *Proc. Natl. Acad. Sci. U.S.A.* **73**, 4334–4338.

Bishop, J. M. (1983). *Annu. Rev. Biochem.* **52**, 301–354.

Blaustein, M. P. (1976). *Fed. Proc., Fed. Am. Soc. Exp. Biol.* **35**, 2574–2578.

Blobel, G., and Dobberstein, B. (1975). *J. Cell Biol.* **67**, 852–862.

Boerner, P., and Racker, E. (1985). *Proc. Natl. Acad. Aci. U.S.A.,* (in press).

Boerner, P., and Saier, M. H., Jr. (1982). *J. Cell. Physiol.* **113**, 240–246.

Boerner, P., and Saier, M. H., Jr. (1985). *J. Cell. Physiol.* **122**, 308–315.

Boerner, P., Resnick, R., and Racker, E. (1985). *Proc. Natl. Acad. Sci. U.S.A.* **82**, 1350–1353.

Bogomolni, R. A., Taylor, M. E., and Stoeckenius, W. (1984). *Proc. Natl. Acad. Sci. U.S.A.* **81**, 5408–5411.

Boheim, G., Hanke, W., Barrantes, F. J., Eibl, H., Sakmann, B., Fels, G., and Maelicke, A. (1981). *Proc. Natl. Acad. Sci. U.S.A.* **78**, 3586–3590.

Boquet, P., and Duflot, E. (1982). *Proc. Natl. Acad. Sci. U.S.A.* **79**, 7614–7618.

Borochov-Neori, H., Yavin, E., and Montal, M. (1984). *Biophys. J.* **45**, 83–85.

Bowman, E. J., Bowman, B. J., and Slayman, C. W. (1981). *J. Biol. Chem.* **256**, 12336–12342.

Boyer, P. D. (1979). *In* "Membrane Bioenergetics" (C. P. Lee, G. Schatz, L. Ernster, eds.), pp. 461–479, Addison-Wesley, Reading, Massachusetts.

Boyer, P. D., Chance, B., Ernster, L., Mitchell, P., Racker, E., and Slater, E. C. (1977). *Annu. Rev. Biochem.* **46**, 955–1026.

Brandolin, G., Meter, C., Defaye, G., Vignais, P. M., and Vignais, P. V. (1974). *FEBS Lett.* **46**, 149–153.

Brandt, D. R., Asano, T., Pedersen, S. E., and Ross, E. M. (1983). *Biochemistry* **22**, 4357–4362.

Braun, S., Raymond, W. E., and Racker, E. (1984). *J. Biol. Chem.* **259**, 2051–2054.

Bremer, E. G., Hakomori, S.-I., Bowen-Pope, D. F., Raines, E., and Ross, R. (1984). *J. Biol. Chem.* **259**, 6818–6825.

Brown, M. S., Anderson, R. G. W., and Goldstein J. L. (1983). *Cell* **32**, 663–667.

Bruni, A., Contessa, A. R., and Luciani, S. (1962). *Biochim. Biophys. Acta* **60**, 301–311.

Büchel, D. E., Gronenborn, B., and Müller-Hill, B. (1980). *Nature (London)* **283**, 541–545.

Bulos, B., and Racker, E. (1968). *J. Biol. Chem.* **243**, 3891–3905.

Burk, D., Woods, M., and Hunter, J. (1967). *J. Natl. Cancer Inst. (U.S.)* **38**, 839–863.

Bürkler, J., and Solioz, M. (1982). *Ann. N. Y. Acad. Sci.* **402**, 422–432.

Burr, J. G., and Rubin, H. (1975). *Cold Spring Harbor Conf. Cell Proliferation* **2**, 807–825.

Cabantchik, Z. I., Knauf, P. A., and Rothstein, A. (1978). *Biochim. Biophys. Acta* **515**, 239–302.

Caffrey, M., and Feigenson, G. W. (1981). *Biochemistry* **20**, 1949–1961.

Campisi, J., Gray, H. E., Pardee, A. B., Dean, M., and Sonenshein, G. E. (1984). *Cell* **36**, 241–247.

Cantley, L. C. (1981). *Curr. Top. Bioenerg.* **11**, 201–237.

Capaldi, R. A., Malatesta, F., and Darley-Usmar, V. M. (1983). *Biochim. Biophys. Acta* **726**, 135–148.

Carafoli, E. (1982). *In* "Membrane Transport of Calcium" (E. Carafoli, ed.), pp. 109–139. Academic Press, London.

Carafoli, E., and Roman, I. (1980). *In* "Molecular Aspects of Medicine" (H. Baum and J. Gergely, eds.), Vol. 3, pp. 295–429. Pergamon, Oxford.

Carafoli, E., and Scarpa, A. (eds.) (1982). "Transport ATPases," Ann. N.Y. Acad. Sci., Vol. 402. N.Y. Acad. Sci., New York.

Carafoli, E., and Zurini, M. (1982). *Biochim. Biophys. Acta* **683,** 279–301.

Carley, W. W., and Racker, E. (1982). *Biochim. Biophys. Acta* **680,** 187–193.

Carlin, C. R., Phillips, P. D., Knowles, B. B., and Cristofalo, V. J. (1983). *Nature (London)* **306,** 617–620.

Caroni, P., Zurini, M., and Clark, A. (1982). *Ann. N.Y. Acad. Sci.* **402,** 402–421.

Carroll, R. C., and Cox, A. C. (1983). *Surv. Synth. Pathol. Res.* **2,** 21–33.

Carroll, R. C., and Racker, E. (1977). *J. Biol. Chem.* **252,** 6981–6990.

Carter-Su, C., and Czech, M. P. (1980). *J. Biol. Chem.* **255,** 10382–10386.

Casey, R. P., Thelen, M., and Azzi, A. (1979). *Biochem. Biophys. Res. Commun.* **87,** 1044–1051.

Casey, R. P., Thelen, M., and Azzi, A. (1980). *J. Biol. Chem.* **255,** 3994–4000.

Cass, C. E., Dahlig, E., Lau, E. Y., Lynch, T. P., and Paterson, A. R. P. (1979). *Cancer Res.* **39,** 1245–1252.

Cassel, D., and Selinger, Z. (1976). *Biochim. Biophys. Acta* **452,** 538–551.

Cassel, D., and Selinger, Z. (1977). *Proc. Natl. Acad. Sci. U.S.A.* **74,** 3307–3311.

Cassel, D., and Selinger, Z. (1978). *Proc. Natl. Acad. Sci. U.S.A.* **75,** 4155–4159.

Cecchini, G., Payne, G. S., and Oxender, D. L. (1979. *J. Supramol. Struct.* **7,** 481–487.

Cerione, R. A., Strulovici, B., Benovic, J. L., Strader, C. D., Caron, M. G., and Lefkowitz, R. J. (1983). *Proc. Natl. Acad. Sci. U.S.A.* **80,** 4899–4903.

Cerione, R. A., Sibley, D. R., Codina, J., Benovic, J. L., Winslow, J., Neer, E. J., Birnbaumer, L., Caron, M. G., and Lefkowitz, R. J. (1984). *J. Biol. Chem.* **259,** 9979–9982.

Changeux, J. P. (1981). *Harvey Lect.* **75,** 85–254.

Changeux, J. P., Heidmann, T., Popot, J.-L., and Sobel, A. (1979). *FEBS Lett.* **105,** 181–187.

Chappell, J. B. (1968). *Br. Med. Bull.* **24,** 150–157.

Cheng, L., and Sacktor, B. (1981). *J. Biol. Chem.* **256,** 1556–1564.

Cheung, W. Y. (1980). *Science* **207,** 19–27.

Cho, T. M., Ge, B. L., Yamoto, C., Smith, A. P., and Loh, H. H. (1983). *Proc. Natl. Acad. Sci. U.S.A.* **80,** 5176–5180.

Christensen, H. N. (1975). "Biological Transport," 2nd ed. Addison-Wesley, Reading, Massachusetts.

Christensen, H. N., Oxender, D. L., Liang, M., and Vatz, K. A. (1965). *J. Biol. Chem.* **240,** 3609–3616.

Christiansen, R. O., Loyter, A., Steensland, H., Saltzgaber, J., and Racker, E. (1969). *J. Biol. Chem.* **244,** 4428–4436.

Churchill, K. A., Holaway, B., and Sze, H. (1983). *Plant Physiol.* **73,** 921–928.

Cidon, S., and Nelson, N. (1983). *J. Biol. Chem.* **258,** 2892–2898.

Cidon, S., Ben-David, H., and Nelson, N. (1983). *J. Biol. Chem.* **258,** 11684–11688.

Citri, Y., and Schramm, M. (1980). *Nature (London)* **287,** 297–300.

Clejan, L., and Beattie, D. S. (1983). *J. Biol. Chem.* **258,** 14271–14275.

Codina, J., Hildebrandt, J. D., Sekura, R. D., Birnbuamer, M., Bryan, J., Manclark, C. R., Iyengar, R., and Birnbaumer, L. (1984). *J. Biol. Chem.* **259,** 5871–5886.

Cohen, G. N., and Rickenberg, H. V. (1955). *C. R. Hebd. Seances Acad. Sci., Ser. D* **240,** 466–468.

Cohn, D. E., and Kaback, H. R. (1980). *Biochemistry* **19,** 4237–4243.

Coin, J. T., and Hinkle, P. C. (1979). *In* "Membrane Bioenergetics" (C. P. Lee, G. Schatz, and L. Ernster, eds.), pp. 405–412. Addison-Wesley, Reading, Massachusetts.

Colca, J. R., McDonald, J. M., Kotagal, N., Patke, C., Fink, C. J., Greider, M. H., Lacy, P. E., and McDaniel, M. L. (1982). *J. Biol. Chem.* **257,** 7223–7228.

Columbini, M. (1980). *Ann. N.Y. Acad. Sci.* **341,** 552–563.

Conover, T. E., Prairie, R. L., and Racker, E. (1963). *J. Biol. Chem.* **238,** 2831–2837.

Conway, E. J. (1951). *Science* **113,** 270–273.

Conzelmann, E., and Sandhoff, K. (1984). *In* "Developmental Neuroscience" (N. Baumann, ed.), pp. 58–71. Karger, Basel.

Cooper, G. M. (1982). *Science* **217,** 801–806.

Cornelius, F., and Skou, J. C. (1984). *Biochim. Biophys. Acta* **772,** 357–373.

Coronado, R., and Miller, C. (1980). *Nature (London)* **288,** 495–497.

Cramer, W. A., Dankert, J. R., and Uratani, Y. (1983). *Biochim. Biophys Acta* **737,** 173–193.

Crane, R. K. (1965). *Fed. Proc, Fed. Am. Soc. Exp. Biol.* **24,** 1000–1006.

Creutz, C. E., Pazoles, C. J., and Pollard, H. B. (1978). *J. Biol. Chem.* **253,** 2858–2866.

Criado, M., Eibl, H., and Barrantes, F. J. (1982). *Biochemistry* **21,** 3622–3629.

Criddle, R. S., Packer, L., and Shieh, P. (1977). *Proc. Natl. Acad. Sci. U.S.A.* **74,** 4306–4310.

Crompton, M., Künzi, M., and Carafoli, E. (1977). *Eur. J. Biochem.* **79,** 549–558.

Crompton, M., Moser, R., Lüdi, H., and Carafoli, E. (1978). *Eur. J. Biochem.* **82,** 25–31.

Cross, R. L. (1981). *Annu Rev. Biochem.* **50,** 681–714.

Cross, R. L., and Nalin, C. M. (1982). *J. Biol. Chem.* **257,** 2874–2881.

Cross, R. L., Grubmeyer, C., and Penefsky, H. S. (1982). *J. Biol. Chem.* **257,** 12101–12105.

Croy, R. G., and Pardee, A. B. (1983). *Proc. Natl. Acad. Sci. U.S.A.* **80,** 4699–4703.

Cullis, P. R., de Kruijff, B., Hope, M. J., Verkleij, A. J., Nayar, R., Forsen, S. B., Tilcock, C., Madden, T. D., and Bolly, M. B. (1983). *In* "Membrane Fluidity in Biology" (R. C. Aloia, ed), Vol. 1, pp. 39–81. Academic Press, New York.

Cunningham, D. D., and Pardee, A. B. (1969). *Proc. Natl. Acad. Sci. U.S.A.* **64,** 1049–1056.

Cuppoletti, J., Jung, C. Y., and Green, F. A. (1981). *J. Biol. Chem.* **256,** 1305–1306.

Curman, B., Klareskog, L., and Peterson, P. A. (1980). *J. Biol. Chem.* **255,** 7820–7826.

Curtis, B. M., and Caterall, W. A. (1984). *Biochemistry* **23,** 2113–2118.

Cushman, S. W., and Wardzala, L. J. (1980). *J. Biol. Chem.* **255,** 4758–4762.

Czech, M. P. (1982). *Cell* **31,** 8–10.

Darmon, A., Zangvill, M., and Cabantchik, Z. I. (1983). *Biochim. Biophys. Acta* **727,** 77–88.

Darszon, A., Vandenberg, C. A., Schönfeld, M., Ellisman, M. H., Spitzer, N. C. and Montal, M. (1980). *Proc. Natl. Acad. Sci. U.S.A.* **77,** 239–243.

Dautry-Varsat, A., Ciechanover, A., and Lodish, H. F. (1983). *Proc. Natl. Acad. Sci. U.S.A.* **80.** 2258–2262.

Deamer, D., and Bangham, A. D. (1976). *Biochim. Biophys. Acta* **443,** 629–634.

Dean, W. L., and Tanford, C. (1977). *J. Biol. Chem.* **252**, 3551–3553.

Dean, W. L., and Tanford, C. (1978). *Biochemistry* **17**, 1683–1690.

Dekowski, S. A., Rybicki, A., and Drickamer, K. (1983). *J. Biol. Chem.* **258**, 2750–2753.

De Larco, J. E., and Todaro, G. J. (1978). *Proc. Natl. Acad. Sci. U.S.A.* **75**, 4001–4005.

de Meis, L. (1982). *Ann. N.Y. Acad. Sci.* **402**, 535–548.

DePont, J. J. H. H. M., van Prooijen-van Eeden, A., and Bonting, S. L. (1978). *Biochim. Biophys. Acta* **508**, 464–477.

Deters, D. W., Racker, E., Nelson, N., and Nelson, H. (1975). *J. Biol. Chem.* **250**, 1041–1047.

Dewey, T. G., and Hammes, G. G. (1981). *J. Biol. Chem.* **256**, 8941–8946.

Diamond, I., Legg, A., Schneider, J. A., and Rozengurt, E. (1978). *J. Biol. Chem.* **253**, 866–871.

Dills, S. S., Apperson, A., Schmidt, M. R., and Saier, M. H., Jr. (1980). *Microbiol. Rev.* **44**, 385–418.

DiMauro, S., Bonilla, E., Lee, C. P., Schotland, A., Scarpa, A., Conn, H. L., and Chance, B. (1976). *J. Neurol. Sci.* **27**, 217.

Dimroth, P. (1981). *J. Biol. Chem.* **256**, 11974–11976.

Dimroth, P., and Hilpert, P. (1984). *16th FEBS Meet.,* Abst., p.71.

Donovan, J. J., Simon, M. I., Draper, R. K., and Montal, M. (1981). *Proc. Natl. Acad. Sci. U.S.A.* **78**, 172–176.

Drachev, L. A., Frolov, V. N., Kaulen, A. D., Liberman, E. A., Ostroumov, S. A., Plakunova, V. G., Semenov, A. Y., and Skulachev, V. P. (1976a). *J. Biol. Chem.* **251**, 7059–7065.

Drachev, L. A., Jasaitis, A. A., Mikelsaar, H., Nemecek, I. B., Semenov, A. Y., Semenova, E. G., Severina, I. I., and Skulachev, V. P. (1976b) *J. Biol. Chem.* **251**, 7077–7082.

Dubinsky, W. P., and Racker, E. (1978). *J. Membr. Biol.* **44**, 25–36.

Dubinsky, W. P., Kandrach, A., and Racker, E. (1979). *In* ''Membrane Bioenergetics'' (C. P. Lee, G. Schatz, and L. Ernster, eds.), pp. 267–280. Addison-Wesley, Reading, Massachusetts.

Ducis, I., and Koepsell, H. (1983). *Biochim. Biophys. Acta* **730**, 119–129.

Dufour, J. P., and Goffeau, A. (1978). *J. Biol. Chem.* **253**, 7026–7032.

Dufour, J. P., and Goffeau, A. (1980). *J. Biol. Chem.* **255**, 10591–10598.

Dunn, S. D., and Futai, M. (1980). *J. Biol. Chem.* **255**, 113–118.

Dunn, S. D., and Heppel, L. A. (1981). *Arch. Biochem. Biophys.* **210**, 421–436.

Dunn, S. D., Heppel, L. A., and Fullmer, C. S. (1980). *J. Biol. Chem.* **255**, 6891–6896.

Enander, K., and Rydström, J. (1982). *J. Biol. Chem.* **257**, 14760–14766.

Enoch, H. G., Fleming, P. J., and Strittmatter, P. (1977). *J. Biol. Chem.* **252**, 5656–5660.

Epstein, M., and Racker, E. (1978). *J. Biol. Chem.* **253**, 6660–6662.

Erni, B., Trachsel, H., Postma, P. W., and Rosenbusch, J. P. (1982). *J. Biol. Chem.* **257**, 13726–13730.

Esch, F. S., and Allison, W. S. (1979). *J. Biol. Chem.* **254**, 10740–10746,

Estes, K., Monfalcone, L., Hammes, S., Holowka, D., and Baird, B. (1985). *Proc. Natl. Acad. Sci. U.S.A.* Submitted.

Eytan, G. D. (1982). *Biochim. Biophys. Acta* **694**, 185–202.

Eytan, G. D., and Broza, R. (1978). *FEBS Lett.* **85**, 175–178.

Eytan, G. D., and Eytan, E. (1980). *J. Biol. Chem.* **255**, 4992–4995.

Eytan, G. D., and Racker, E. (1977). *J. Biol. Chem.* **252**, 3208–3213.

Eytan, G. D., Matheson, M. J., and Racker, E. (1975). *FEBS Lett.* **57**, 121–125.

Eytan, G. D., Matheson, M. J., and Racker, E. (1976). *J. Biol. Chem.* **251**, 6831–6837.

Eytan, G. D., Gad, A., Broza, R., and Eytan, E. (1979). *In* "Membrane Bioenergetics" (C. P. Lee, G. Schatz, and L. Ernster, eds.) pp. 308–318. Addison-Wesley, Reading, Massachusetts.

Fagan, J. B., and Racker, E. (1978). *Cancer Res.* **38**, 749–758.

Fairbanks, G., Steck, T. L., and Wallach, D. F. H. (1971). *Biochemistry* **10**, 2606–2617.

Fairclough, P., Malathi, P., Preiser, H., and Crane, R. K. (1979). *Biochim. Biophys. Acta* **553**, 295–306.

Fairclough, R. H., Finer-Moore, J., Love, R. A., Kristofferson, D., Desmeules, P. J., and Stroud, R. M. (1983) *Cold Spring Harbor Symp. Quant. Biol.* **48**, 9–20.

Faller, L., Jackson, R., Malinowska, D., Mukidjam, E., Rabon, E., Saccomani, G., Sachs, G., and Smolka, A. (1982). *Ann. N. Y. Acad. Sci.* **402**, 146–163.

Fang, J.-K., Jacobs, J. W., Kanner, B. I., Racker, E., and Bradshaw, R. A. (1984). *Proc. Natl. Acad. Sci. U.S.A.* **81**, 6603–6607.

Feldman, R. I., and Sigman, D. S. (1982). *J. Biol. Chem.* **257**, 1676–1683.

Ferguson, S. J., Lloyd, W. J., Lyons, M. H., and Radda, G. K. (1975). *Eur. J. Biochem.* **54**, 117–126.

Ferguson-Miller, S., Brautigan, D. L., and Margoliash, E. (1978). *J. Biol. Chem.* **253**, 149–159.

Fessenden-Raden, J. M. (1969). *J. Biol. Chem.* **244**, 6662–6667.

Fidler, I. J., Sone, S., Fogler, W. E., and Barnes, Z. L. (1981). *Proc. Natl. Acad. Sci. U.S.A.* **78**, 1680–1684.

Fillingame, R. H. (1981). *Curr. Top. Bioenerg.* **11**, 35–106.

Fishkes, H., and Rŭdnick, G. (1982). *J. Biol. Chem.* **257**, 5671–5677.

Fleischer, S., and McIntyre, J. O. (1982). *Ann. N.Y. Acad. Sci.* **402**, 558–560.

Flügge, U. I., and Heldt, H. W. (1981). *Biochim. Biophys. Acta* **638**, 296–304.

Flügge, U. I., Gerber, J., and Heldt, H. W. (1983). *Biochim. Biophys Acta* **725**, 229–237.

Folch-Pi, J., and Stoffyn, P. J. (1972). *Ann. N.Y. Acad. Sci.* **195**, 86–107.

Forbes, D. J., Kirschner, M. W., and Newport, J. W. (1983). *Cell* **34**, 13–23.

Forbush, B., III, Kaplan, J. H., and Hoffman, J. F. (1978). *Biochemistry* **17**, 3667–3676.

Forgac, M., Cantley, L., Wiedenmann, B., Altstiel, L., and Branton, D. (1983). *Proc. Natl. Acad. Sci. U.S.A.* **80**, 1300–1303.

Forstner, G. G. (1971). *Biochem. J.* **121**, 781–789.

Foster, D. L., and Fillingame, R. H. (1982). *J. Biol. Chem.* **257**, 2009–2015.

Foster, D. L., Garcia, M. L., Newman, M. J., Patel, L., and Kaback, H. R. (1982). *Biochemistry* **21**, 5634–5638.

Foster, D. O., and Pardee, A. B. (1969). *J. Biol. Chem.* **244**, 2675–2681.

Fox, C. F., and Kennedy, E. P. (1965). *Proc. Natl. Acad. Sci. U.S.A.* **54**, 891–899.

Fraley, R., Subramani, S., Berg, P., and Papahadjopoulos, D. (1980). *J. Biol. Chem.* **255**, 10431–10435.

Franzusoff, A. J., and Cirillo, V. P. (1983). *Biochim. Biophys Acta* **734**, 153–159.

Freitag, H., Neupert, W., and Benz, R. (1982). *Eur. J. Biochem* **123**, 629–636.

Frelin, C., Vigne, P., and Lazdunski, M. (1983). *J. Biol. Chem.* **258**, 6272–6276.

Fujita, N., Nelson, N., Fox, T. D., Claudio, T., Lindstrom, J., Riezman, H., and Hess, G. P. (1985). *Science* (submitted).

Fung, B. K.-K. (1983) *J. Biol. Chem.* **258**, 10495–10502.

Futai, M., and Kanazawa, H. (1980). *Curr. Top. Bioenerg.* **10**, 181–215.

Futai, M., and Kanazawa, H. (1983). *Microbiol. Rev.* **47**, 285–312.

Gad, A. E., Broza, R., and Eytan, G. D. (1979). *FEBS Lett.* **102**, 230–234.

Galjaard, H. (1980). "Genetic Metabolic Diseases; Early Diagnosis and Prenatal Analysis." Elsevier/North-Holland Biomedical Press, Amsterdam.

Galjaard, H. (1982). *Methods Cell Biol.* **26**, 241–268.

Galjaard, H., and Reuser, A. J. J. (1984). *In* "Lysosomes in Biology and Pathology (J. T., Dingle, R. T. Dean, and W. Sly, eds.), pp. 315–345. Elsevier, Amsterdam.

Gallico, G. G., III, O'Connor, N. E., Compton, C. C., Kehinde, O., and Green, H. (1984). *N. Engl. J. Med.* **311**, 448–451.

Gammon, M. T., and Isselbacher, K. J. (1976). *J. Cell Physiol.* **89**, 759–764.

Ganser, A. L., and Forte, J. G. (1973). *Biochim. Biophys. Acta* **307**, 169–180.

Garavito, R. M., and Rosenbusch, J. P. (1980). *J. Cell Biol.* **86**, 327–329.

Garavito, R. M., Jenkins, J. A., Neuhaus, J. M., Pugsley, A. P., and Rosenbusch, J. P. (1982). *Ann. Microbiol.* (Paris) **133**, 37–41.

Garber, M. P., and Steponkus, P. L. (1974). *J. Cell Biol.* **63**, 24–34.

Gasko, O. D., Knowles, A. F., Shertzer H. G., Suolinna, E.-M., and Racker, E. (1976). *Anal. Biochem.* **72**, 57–65.

Gatt, S., and Racker, E. (1959). *J. Biol. Chem.* **234**, 1015–1023.

Gay, J. N., and Walker, J. E. (1981). *Nucleic Acids Res.* **9**, 3919–3926.

Gerber, G. E., Gray, C. P., Wildenauer, D., and Khorana, H. G. (1977). *Proc. Natl. Acad. Sci. U.S.A.* **74**, 5426–5430.

German, D. C., Bloch, C. A., and Kredich, N. M. (1983). *J. Biol. Chem.* **258**, 10997–11003.

Ghanotakis, D. F., Topper, J. N., Babcock, G. T., and Yocum, C. F. (1984). *FEBS Lett.* **170**, 169–173.

Gibson, F. (1983). *Biochem. Soc. Trans.* **11**, 229–240.

Gibson, Q. H., Greenwood, C., Wharton, D. C., and Palmer, G. (1965). *J. Biol. Chem.* **240**, 888–894.

Gietzen, K., Tejčka, M., and Wolf, U. J. (1980). *Biochem. J.* **189**, 81–88.

Gillies, R. J. (1981). *In* "The Transformed Cell" (I. L. Cameron and T. B. Pool, eds.), pp. 347–395. Academic Press, New York.

Gilmore, T., DeClue, J. E., and Martin, G. S. (1985). *Cell* **40**, 609–618.

Goffeau, A., and Slayman, C. W. (1981). *Biochim. Biophys. Acta* **639**, 197–223.

Goldin, S. G., and Rhoden, V. (1978). *J. Biol. Chem.* **253**, 2575–2583.

Goldin, S. M. (1979). *In* "Na, K-ATPase Structure and Kinetics" (J. C. Skou and J. G. Norby, eds.), pp. 69–85. Academic Press, New York.

Goldin, S. M., and Tong, S. W. (1974). *J. Biol. Chem.* **249**, 5907–5915.

Goldin, S. M., Rhoden, V., and Hess, E. J. (1980). *Proc. Natl. Acad. Sci. U.S.A.* **77**, 6884–6888.

Goldstein, J. L., Anderson, R. G. W., and Brown, M. S. (1979). *Nature (London)* **279,** 679–685.

Gómez-Puyou, A., Gómez-Puyou, M. T., and Ernster, L. (1979). *Biochim. Biophys. Acta* **547,** 252–257.

Gottesman, M. M., and Cabral, F. (1981). *Biochemistry* **20,** 1659–1665.

Gould-Fogerite, S., and Mannino, R. J. (1985). *Anal. Biochem.* **148,** 15–25.

Gräber, P., Rögner, M., Buchwald, H.-E., Samoray, D., and Hauska, G. (1982). *FEBS Lett.* **145,** 35–40.

Grabo, M. (1982). Ph.D. Thesis, University of Basel, Switzerland (under supervision of Dr. J. Rosenbusch).

Green, N. M., Allen, G., and Hebdon, G. M. (1980). *Ann. N. Y. Acad. Sci.* **358,** 149–158.

Groudine, M., and Weintraub, H. (1980). *Proc. Natl. Acad. Sci. U.S.A.* **77,** 5351–5354.

Guffanti, A. A., Cohn, D. E., Kaback, H. R., and Krulwich, T. A. (1981). *Proc. Natl. Acad. Sci. U.S.A.* **78,** 1481–1484.

Guidotti, G. G., Borghetti, A. F., and Gazzola, G. C. (1978). *Biochim. Biophys. Acta* **515,** 329–366.

Gwynn, G. J., and Costa, E. (1982). *Proc. Natl. Acad. Sci. U.S.A.* **79,** 690–694.

Haaker, H., and Racker, E. (1979). *J. Biol. Chem.* **254,** 6598–6602.

Hackenberg, H., and Klingenberg, M. (1980). *Biochemistry* **19,** 548–555.

Hakomori, S. (1981). *Annu. Rev. Biochem.* **50,** 733–764.

Halestrap, A. P. (1976). *Biochem. J.* **156,** 193–207.

Hall, C., and Ruoho, A. (1980). *Proc. Natl. Acad. Sci. U.S.A.* **77,** 4529–4533.

Hamilton, R. T., and Nilsen-Hamilton, M. (1978). *J. Biol. Chem.* **253,** 8247–8256.

Hammill, O. P., Marty, A., Neher, E., Sakmann, B., and Sigworth, F. J. (1981). *Pfluegers Arch.* **391,** 85–100.

Happe, M., Teather, R. M., Overath, P., Knobling, A., and Oesterhelt, D. (1977). *Biochim. Biophys. Acta* **465,** 415–420.

Harold, F. M., and Levin, E. (1974). *J. Bacteriol.* **117,** 1141–1148.

Harris, D. A., and Crofts, A. R. (1978). *Biochim. Biophys. Acta* **502,** 87–102.

Hartshorne, R. P., and Catterall, W. A. (1984). *J. Biol. Chem.* **259,** 1667–1675.

Hartshorne, R. P., Keller, B. U., Talvenheimo, J. A., Catterall, W. A., and Montal, M. (1985). *Proc. Natl. Acad. Sci. U.S.A.* **82,** 240–244.

Hasselbach, W., and Waas, W. (1982). *Ann. N.Y. Acad. Sci.* **402,** 459–469.

Hatefi, Y., Haavik, A. G., Fowler, L. R., and Griffiths, D. E. (1962). *J. Biol. Chem.* **237,** 2661–2669.

Hauska, G., Samoray, D., Orlich, G., and Nelson, N. (1980). *Eur. J. Biochem.* **111,** 535–543.

Hauska, G., Hurt, E., Gabellini, N., and Lockau, W. (1983). *Biochim. Biophys. Acta* **726,** 97–133.

Hayflick, L., and Moorhead, P. S. (1961). *Exp. Cell Res.* **25,** 585–621.

Haywood, A. M. (1983). *In* "Liposomes Letters" (A. D. Bangham, ed.) pp. 277–287. Academic Press, New York.

Heefner, D. L., and Harold, F. M. (1982). *Proc. Natl. Acad. Sci. U.S.A.* **79,** 2798–2802.

Heefner, D. L., Kobayashi, H., and Harold, F. M. (1980). *J. Biol. Chem.* **255,** 11403–11407.

Heidmann, T., Sobel, A., Popot, J.-L., and Changeux, J.-P. (1980). *Eur. J. Biochem.* **110,** 35–55.

Helenius, A., and Simons, K. (1975). *Biochim. Biophys. Acta* **415,** 29–79.

Helenius, A., Fries, E., and Kartenbeck, J. (1977). *J. Cell Biol.* **75,** 866–880.

Helenius, A., McCaslin, D. R., Fries, E., and Tanford, C. (1979). *In* "Methods in Enzymology" (S. Fleischer and L. Packer, eds.) Vol. 56, pp. 734–749. Academic Press, New York.

Helenius, A., Mellman, I., Wall, D., and Hubbard, A. (1983). *Trends Biochem. Sci.* **8,** 245–250.

Henderson, P. J. F., and Lardy, H. A. (1970). *J. Biol. Chem.* **245,** 1319–1326.

Henderson, P. J. F., Kagawa, Y., and Hirata, H. (1983). *Biochim. Biophys Acta* **732,** 204–209.

Henderson, R., and Unwin, P. N. T. (1975). *Nature (London)* **257,** 28–32.

Henderson, R., Capaldi, R. A., and Leigh, J. S. (1977). *J. Mol. Biol.* **112,** 631–648.

Hess, G. P., and Andrews, J. P. (1977). *Proc. Natl. Acad. Sci. U.S.A.* **74,** 482–486.

Hidalgo, C., Petrucci, D. A., and Vergara, C. (1982). *J. Biol. Chem.* **257,** 208–216.

Hilden, S., Rhee, H. M., and Hokin, L. E. (1974). *J. Biol. Chem.* **249,** 7432–7440.

Hinkle, P. (1973). *Fed. Proc., Fed. Am. Soc. Exp. Biol.* **32,** 1988–1992.

Hinkle, P. C., Kim, J. J., and Racker, E. (1972). *J. Biol. Chem.* **247,** 1338–1339.

Hinkle, P. C., Sogin, D. C., Wheeler, T. J., and Telford, J. N. (1979). *In* "Function and Molecular Aspects of Biomembrane Transport" (E. Quagliariello, F. Palmieri, S. Papa, and M. Klingenberg, eds.), pp. 487–494. Elsevier/North-Holland Biomedical Press, Amsterdam.

Hinnen, R., Miyamoto, H., and Racker, E. (1979). *J. Membr. Biol.* **49,** 309–324.

Hirata, H., Kambe, T., and Kagawa, Y. (1984). *J. Biol. Chem.* **259,** 10653–10656.

Hoffman, F. M. (1979). *Biochem. Biophys. Res. Commun.* **86,** 988–994.

Hoffman, R. M. (1984). *Biochim. Biophys. Acta* **738,** 49–87.

Höjeberg, B., Lind, C., and Khorana, H. G. (1982). *J. Biol. Chem.* **257,** 1690–1694.

Hokin, L. E., and Dixon, J. F. (1979). *In* "Na, K-ATPase Structure and Kinetics" (J. C. Skou and J. G. Norby, eds.), pp. 47–67. Academic Press, New York.

Hollenberg, M. D., Goren, H. J., Hanif, K., and Lederis, K. (1983). *Trends Pharmacol. Sci.* **4,** 310–312.

Holley, R. W. (1975). *Nature (London)* **258,** 487–490.

Holley, R. W., and Kiernan, J. A. (1968). *Proc. Natl. Acad. Sci. U.S.A.* **60,** 300–304.

Holowka, D. A., and Hammes, G. G. (1977). *Biochemistry* **16,** 5538–5545.

Hong, K., Düsgünes, N., and Papahadjopoulos, D. (1981). *J. Biol. Chem.* **256,** 3641–3644.

Hopfer, U., Nelson, K., Perrotto, J., and Isselbacher, K. J. (1973) *J. Biol. Chem.* **248,** 25–32.

Hoppe, J., Friedl, P., Schairer, H. U., Sebald, W., von Meyenburg, K., and Jørgensen, B. B. (1983). *EMBO J.* **2,** 105–110.

Horstman, L. L., and Racker, E. (1970). *J. Biol. Chem.* **245,** 1336–1344.

Huang, K.-S., Bayley, H., Liao, M.-J., London, E., and Khorana, H. G. (1981). *J. Biol. Chem.* **256,** 3802–3809.

Hüdig, H., and Drews, G. (1984). *Biochim. Biophys. Acta* **765,** 171–177.

Huganir, R. L., and Racker, E. (1980). *J. Supramol. Struct.* **14,** 13–19.

Huganir, R. L., and Racker, E. (1982). *J. Biol. Chem.* **257,** 9372–9378.

Huganir, R. L., Schell, M. A., and Racker, E. (1979). *FEBS Lett.* **108,** 155–160.

Huganir, R. L., Coronado, R., Silverman, D. H., and Racker, E. (1981). *Int. Biophys. Congr., 7th, 1981, Biochem. Pan-Am. Congr. 3rd, 1981*, p. 258.

Hughes, J. B., Joshi, S., Murfitt, R. R., and Sanadi, D. R. (1979). *In* "Membrane Bioenergetics" (C. P. Lee, G. Schatz, and L. Ernster, eds.), pp. 81–95. Addison-Wesley, Reading, Massachusetts.

Hughes, J. B., Joshi, S., Torok, K., and Sanadi, D. R. (1982). *J. Bioenerg. Biomembr.* **14**, 287–295.

Hugues, M., Romey, G., Duval, D., Vincent, J. P., and Lazdunski, M. (1982). *Proc. Natl. Acad. Sci. U.S.A.* **79**, 1308–1312.

Hundal, T., and Ernster, L. (1979). *In* "Membrane Bioenergetics" (C. P. Lee, G. Schatz, and L. Ernster, eds.), pp. 429–445. Addison-Wesley, Reading, Massachusetts.

Hunter, T. (1984). *Sci. Am.* **251**, 70–79.

Hunter, T. (1985). *Trends Biochem. Sci.* **10**, 275–280.

Ingebretsen, O. C., Terland, O., and Flatmark, T. (1980). *Biochim. Biophys. Acta* **628**, 182–189.

Inman, W. H., and Colowick, S. P. (1985). *Proc. Natl. Acad. Sci. U.S.A.* **82**, 1346–1349.

Isselbacher, K. J. (1972). *Proc. Natl. Acad. Sci. U.S.A.* **69**, 585–589.

Iwata, K. K., Fryling, C. M., Knott, W. B., and Todaro, G. J. (1983). *Fed. Proc., Fed. Am. Soc. Exp. Biol.* **42**, 1833.

Jacobs, E. E., and Sanadi, D. R. (1960). *J. Biol. Chem.* **235**, 531–534.

Jardetzky, O. (1966). *Nature (London)* **211**, 969–970.

Jarvis, S. M., McBride, D., and Young, J. D. (1982). *J. Physiol. (London)* **324**, 31–46.

Jasàitis, A. A., Nemeček, I. B., Severina, I. I., Skulachev, V. P., and Smirnova, S. M. (1972). *Biochim. Biophys. Acta* **275**, 485–490.

Jeng, A. Y., and Shamoo, A. E. (1980). *J. Biol. Chem.* **255**, 6897–6903.

Jennings, M. L., and Adams-Lackey, M. (1982). *J. Biol. Chem.* **257**, 12866–12871.

Jeon, K. W., Lorch, I. J., and Danielli, J. F. (1970). *Science* **167**, 1626–1627.

Johannsson, A., Smith, G. A., and Metcalfe, J. C. (1981). *Biochim. Biophys. Acta* **641**, 416–421.

John, P., and Whatley, F. R. (1977). *Biochim. Biophys. Acta* **463**, 129–153.

Johnson, J. H., Belt, J. A., Dubinsky, W. P., Zimniak, A., and Racker, E. (1980). *Biochemistry* **19**, 3835–3840.

Johnson, M. J. (1941). *Science* **94**, 200–202.

Johnson, P. A., and Johnstone, R. M. (1982). *Membr. Biochem.* **4**, 189–218.

Johnson, R. G., Carty, S., and Scarpa, A. (1982). *Fed. Proc., Fed. Am. Soc. Exp. Biol.* **41**, 2746–2745.

Johnstone, R. M., and Bardin, C. (1976). *J. Cell Physiol.* **89**, 801–804.

Jones, M. N., and Nickson, J. K. (1981). *Biochim. Biophys. Acta* **650**, 1–20.

Jones, T. H. D., and Kennedy, E. P. (1969). *J. Biol. Chem.* **244**, 5981–5987.

Jørgensen, P. L. (1974). *Qu. Rev. Biophys.* **7**, 239–274.

Jørgensen, P. L., and Anner, B. M. (1979). *Biochim. Biophys. Acta* **555**, 485–492.

Jørgensen, P. L., Skriver, E., Hebert, H., and Maunsbach, A. B. (1982). *Ann. N.Y. Acad. Sci.* **402**, 207–225.

Julliard, J. H., and Gautheron, D. C. (1978). *Biochim. Biophys Acta* **503**, 223–237.

Kaback, H. R. (1983). *J. Membr. Biol.* **76**, 95–112.

Kagawa, Y. (1972). *Biochim. Biophys. Acta* **265**, 297–338.

Kagawa, Y. (1978). *Biochim. Biophys. Acta* **505**, 45–93.

Kagawa, Y. (1980). *J. Membr. Biol.* **55**, 1–8.

Kagawa, Y., and Racker, E. (1966). *J. Biol. Chem.* **241**, 2461–2482.

Kagawa, Y., and Racker, E. (1971). *J. Biol. Chem.* **246**, 5477–5487.

Kagawa, Y., Sone, N., Hirata, H., Yoshida, M., Rögner, M., and Ohta, S. (1979). *In* "Membrane Bioenergetics" (C. P. Lee, G. Schatz, and L. Ernster, eds.), pp. 177–188. Addison-Wesley, Reading, Massachusetts.

Kagawa, Y., Kandrach, A., and Racker, E. (1973a). *J. Biol. Chem.* **248**, 676–684.

Kagawa, Y., Johnson, L. W., and Racker, E. (1973b). *Biochem. Biophys. Res. Commun.* **50**, 245–251.

Kagawa, Y., Ide, C., Hamamoto, T., Rögner, M., and Sone, N. (1982). *Cell Surf. Rev.* **8**, 137–160.

Kamienietzky, A., and Nelson, N. (1975). *Plant Physiol.* **55**, 282–287.

Kanazawa, H., and Futai, M. (1982). *Ann. N.Y. Acad. Sci.* **402**, 45–64.

Kanazawa, H., Horiuchi, Y., Takagi, M., Ishino, Y., and Futai, M. (1980). *J. Biochem. (Tokyo)* **88**, 695–703.

Kanner, B. I. (1978). *FEBS Lett.* **89**, 47–50.

Kanner, B. I., and Sharon, I. (1978). *FEBS Lett.* **94**, 245–248.

Kanner, B. I., Serrano, R., Kandrach, M. A., and Racker, E. (1976). *Biochem. Biophys. Res. Commun.* **69**, 1050–1056.

Kantham, B. C. L., Hughes, J. B., Pringle, M. J., and Sanadi, D. R. (1984) *J. Biol. Chem.* **259**, 10627–10632.

Kantor, H. L., and Prestegard, J. H. (1975). *Biochemistry* **14**, 1790–1795.

Karlin, A. (1980). *In* "The Cell Surface and Neuronal Function" (G. Poste, G. Nicolson, and C. Cotman, eds.), pp. 191–260. Elsevier/North-Holland Biomedical Press, New York.

Karlish, S. J. D., and Pick, U. (1981). *J. Physiol.* (London) **312**, 505–529.

Karlish, S. J. D., and Stein, W. D. (1982). *Ann. N.Y. Acad. Sci.* **402**, 226–238.

Kasahara, M., and Hinkle, P. C. (1976). *Proc. Natl. Acad. Sci. U.S.A.* **73**, 396–400.

Kasuga, M., Zick, Y., Blithe, D. L., Crettaz, M., and Kahn, C. R. (1982). *Nature (London)* **298**, 667–669.

Kay, M. M. B., Goodman, S. R., Sorensen, K., Whitfield, C. F., Wong, P., Zaki, L., and Rudloff, V. (1983). *Proc. Natl. Acad. Sci. U.S.A.* **80**, 1631–1635.

Kessar, P., and Crompton, M. (1981). *Biochem. J.* **200**, 379–388.

Khorana, H. G., Gerber, G. E., Herlihy, W. C., Gray, C. P., Anderegg, R. J., Nihei, K., and Biemann, K. (1979). *Proc. Natl. Acad. Sci. U.S.A.* **76**, 5046–5050.

Kilberg, M. S. (1982). *J. Membr. Biol.* **69**, 1–12.

Kilian, P. L., Dunlap, C. R., Mueller, P., Schell, M. A., Huganir, R. L., and Racker, E. (1980). *Biochem. Biophys. Res. Commun.* **93**, 409–414.

Kimball, E. S., Bohn, W. H., Cockley, K. D., Warren, T. C., and Sherwin, S. A. (1984). *Cancer Res.* **44**, 3613–3619.

King, T. E. (1963). *J. Biol.* **238**, 4037–4051.

King, T. E. (1981). *In* "Chemiosmotic Proton Circuits in Biological Membranes (V. P. Skulachev and P. C. Hinkle, eds.), pp. 147–159. Addison-Wesley, Reading, Massachusetts.

Kirchberger, M. A., and Tada, M. (1976). *J. Biol. Chem.* **251**, 725–729.

Klausner, R. D., Bridges, K., Tsunoo, H., Blumenthal, R., Weinstein, J. N., and Ashwell, G. (1980). *Proc. Natl. Acad. Sci. U.S.A.* **77**, 5087–5091.

Klausner, R. D., Ashwell, G., van Renswoude, J., Harford, J. B., and Bridges, K. R. (1983). *Proc. Natl. Acad. Sci. U.S.A.* **80**, 2263–2266.

Klausner, R. D., Renswoude, J. V., Blumenthal, R., and Rivnay, B. (1984). *In* "Molecular and Chemical Characterization of Membrane Receptors" (J. C. Venter and L. C. Harrison, eds.), Vol. 3, pp. 209–239. Alan R. Liss, Inc., New York.

Klingenberg, M. (1976). *In* "The Enzymes of Biological Membranes: Membrane Transport" (A. R. Martonosi, ed.), Vol. 3, pp. 383–438. Plenum, New York.

Knoth, J., Zallakian, M., and Njus, D. (1982). *Fed. Proc., Fed. Am. Soc. Exp. Biol.* **41**, 2742–2745.

Knowles, A. F., and Racker, E. (1975a). *J. Biol. Chem.* **250**, 1949–1951.

Knowles, A. F., and Racker, E. (1975b). *J. Biol. Chem.* **250**, 3538–3544.

Knowles, A. F., Guillory, R. J., and Racker, E. (1971). *J. Biol. Chem.* **246**, 2672–2679.

Knowles, A. F., Kandrach, A., Racker, E., and Khorana, H. G. (1975). *J. Biol. Chem.* **250**, 1809–1813.

Knowles, A. F., Eytan, E., and Racker, E. (1976). *J. Biol. Chem.* **251**, 5161–5165.

Knowles, A., Zimniak, P., Alfonzo, M., Zimniak, A., and Racker, E. (1980). *J. Membr. Biol.* **55**, 233–239.

Kobayashi, H., Van Brunt, J., and Harold, F. M. (1978). *J. Biol. Chem.* **253**, 2085–2092.

Koepsell, H., Menuhr, H., Ducis, I., and Wissmüller, T. F. (1983). *J. Biol. Chem.* **258**, 1888–1894.

Köhne, W., Deuticke, B., and Haest, C. W. M. (1983). *Biochim. Biophys Acta* **730**, 139–150.

König, B., and Sandermann, H., Jr. (1982). *FEBS Lett.* **147**, 31–34.

Kono, T., Robinson, F. W., Blevins, T. L., and Ezaki, O. (1982). *J. Biol. Chem.* **257**, 10942–10947.

Kováč, L., Lachowicz, T. M., and Slonimski, P. P. (1967). *Science* **158**, 1564–1567.

Kozlov, I. A. (1981). *In* "Chemiosmotic Proton Circuits in Biological Membranes" (V. P. Skulachev and P. C. Hinkle, eds.), pp. 407–420. Addison-Wesley, Reading, Massachusetts.

Krämer, R., and Klingenberg, M. (1979). *Biochemistry* **18**, 4209–4215.

Kris, R. M., Lax, I., Gullick, Waterfield, M. D., Ullrich, A., Fridkin, M., and Schlessinger, J. (1985). *Cell* **40**, 619–625.

Krueger, B. K., Worley, J. F., III, and French, R. J. (1983). *Nature (London)* **303**, 172–175.

Krulwich, T. A. (1983). *Biochim. Biophys. Acta* **726**, 245–264.

Kundig, W., and Roseman, S. (1971). *J. Biol. Chem.* **246**, 1407–1418.

Kundig, W., Ghosh, S., and Roseman, S. (1964). *Proc. Natl. Acad. Sci. U.S.A.* **52**, 1067–1074.

Kuriki, Y., and Racker, E. (1976). *Biochemistry* **15**, 4951–4956.

Kyte, J. (1975). *J. Biol. Chem.* **250**, 7443–7449.

LaBelle, E. F. (1984). *Biochim. Biophys. Acta* **770**, 79–92.

LaBelle, E. F., and Lee, S. O. (1982). *Biochemistry* **21**, 2693–2697.

Land, H., Parada, L. F., and Weinberg, R. A. (1983). *Nature (London)* **304**, 596–602.

Lang, D. R., and Racker, E. (1974). *Biochim. Biophys. Acta* **333**, 180–186.

Langridge-Smith, J. E., Field, M., and Dubinsky, W. P. (1984). *Biochim. Biophys. Acta* **777**, 84–92.

Lardy, H. A. (1980). *Pharmacol. Ther.* **11**, 649–660.

Lardy, H. A., Johnson, D., and McMurray, W. C. (1958). *Arch. Biochem. Biophys.* **78**, 587–597.

Lee, C. P., Schatz, G., and Ernster, L., eds. (1979) "Membrane Bioenergetics." Addison-Wesley, Reading, Massachusetts.

Lee, S.-H. Kalra, V. K., and Brodie, A. F. (1979). *J. Biol. Chem.* **254**, 6861–6864.

Lefkowitz, R. J., Stadel, J. M., and Caron, M. G. (1983). *Annu. Rev. Biochem.* **52**, 159–186.

Lefkowitz, R. J., Caron, M. G., and Stiles, G. L. (1984). *New Engl. J. Med.* **310**, 1570–1579.

Lehninger, A. L., Wadkins, C. L., and Remmert, L. F. (1959). *In* "Regulation of Cell Metabolism" (G. E. W. Wolstenholme and C. M. O'Connor, eds.), Ciba Found. Symp. pp. 130–149. Churchill, London.

Lehninger, A. L., Reynafarje, B., Alexandre, A., and Villalobo, A. (1979). *In* "Membrane Bioenergetics" (C. P. Lee, G. Schatz, and L. Ernster, eds.), pp. 393–404. Addison-Wesley, Reading, Massachusetts.

Lemire, B. D., Robinson, J. J., Bradley, R. D., Scraba, D. G., and Weiner, J. H. (1983). *J. Bacteriol.* **155**, 391–397.

Leonard, J. E., and Saier, M. H., Jr. (1983). *J. Biol. Chem.* **258**, 10757–10760.

Le Peuch, C. J., Haiech, J., and Demaille, J. G. (1979). *Biochemistry* **18**, 5150–5157.

Le Peuch, C. J., Ballester, R., and Rosen, O. M. (1983). *Proc. Natl. Acad. Sci. U.S.A.* **80**, 6858–6862.

Leung, K. H., and Hinkle, P. C. (1975). *J. Biol. Chem.* **250**, 8467–8471.

Lever, J. E. (1977). *J. Biol. Chem.* **252**, 1990–1997.

Levine, A. E., Hamilton, D. A., Yoeman, L. C., Busch, H., and Brattain, M. G. (1984). *Biochem. Biophys. Res. Commun.* **119**, 76–82.

Levitan, I. B., Lemos, J. R., and Novak-Hofer, I. (1983). *Trends NeuroSci.* **6**, 496–499.

Liao, M.-J., London, E., and Khorana, H. G. (1983). *J. Biol. Chem.* **258**, 9949–9955.

Lichtenberg, D., Robson, R. J., and Dennis, E. A. (1983). *Biochim. Biophys. Acta* **737**, 285–304.

Lien, S., and Racker, E. (1971). *J. Biol. Chem.* **246**, 4298–4307.

Lind, C., Höjeberg, B., and Khorana, H. G. (1981). *J. Biol. Chem.* **256**, 8298–8305.

Linden, M., and Gellerfors, P. (1983). *Biochim. Biophys. Acta* **736**, 125–129.

Ling, V., Kartner, N., Sudo, T., Siminovitch, L., and Riordan, J. R. (1983). *Cancer Treat. Rep.* **67**, 869–874.

Littlefield, J. W. (1982). *Science* **218**, 214–216.

Lo, M. M. S., Strittmatter, S. M., and Snyder, S. H. (1982). *Proc. Natl. Acad. Sci. U.S.A.* **79**, 680–684.

Lowe, A. G., and Lambert, A. (1983). *Biochim. Biophys. Acta* **694**, 353–374.

Lowry, O. H., Rosebrough, N. J., Farr, A. L., and Randal, R. J. (1951). *J. Biol. Chem.* **193**, 265–275.

Loyter, A., and Volsky, D. J. (1979). *In* "Membrane Bioenergetics (C. P. Lee, G. Schatz, and L. Ernster, eds.), pp. 319–330. Addison-Wesley, Reading, Massachusetts.

Loyter, A., Saltzgaber, J., Steensland, H., and Racker, E. (1969). *Ann. N.Y. Acad. Sci.* **147**, 846–848.

Ludwig, B. (1980). *Biochim. Biophys. Acta* **594**, 177–189.

Ludwig, B., and Schatz, G. (1980). *Proc. Natl. Acad. Sci. U.S.A.* **77**, 196–200.

Luft, R., Ikkos, D., Palmieri, G., Ernster, L., and Afzelius, B. (1962). *J. Clin. Invest.* **41,** 1776–1804.

Lukacovic, M. F., Feinstein, M. B., Sha'afi, R. I., and Perrie, S. (1981). *Biochemistry* **20,** 3145–3151.

Luria, S. E., and Suit, J. L. (1982). *In* "Membranes and Transport" (A. N. Martonosi, ed.), Vol. 2, pp. 279–284. Plenum, New York.

Lynen, F. (1941). *Justus Liebigs Ann. Chem.* **246,** 120–141.

McCara, I. G., Marinetti, G. V., and Balduzzi, P. C. (1984). *Proc. Natl. Acad. Sci. U.S.A.* **81,** 2728–2732.

McCarty, R. E. (1979). *Annu Rev. Plant Physiol.* **30,** 79–104.

McCarty, R. E., and Fagan, J. (1973). *Biochemistry* **12,** 1503–1507.

McCarty, R. E., and Racker, E. (1966). *Brookhaven Symp. Biol.* **19,** 202–214.

McCormick, J. I., Tsang, D., and Johnstone, R. M. (1984). *Arch. Biochem. Biophys.* **231,** 355–365.

McCoy, G. D., Resch, R. C., and Racker, E. (1976). *Cancer Res.* **36,** 3339–3345.

McDonald, J. M., Chan, K.-M., Goewert, R. R., Mooney, R. A., and Pershadsingh, H. A. (1982). *Ann. N.Y. Acad. Sci.* **402,** 381–401.

MacDonald, R. E., Greene, R. V., Clark, R. D., and Lindley, E. V. (1979). *J. Biol. Chem.* **254,** 11831–11838.

MacDonald, R. I., Oku, N., and MacDonald, R. C. (1983). *In* "Liposomes Letters" (A. D. Bangham, ed.), pp. 63–72. Academic Press, New York.

McDonough, A. A., Hiatt, A., and Edelman, I. S. (1982). *J. Membr. Biol.* **69,** 13–22.

MacLennan, D. H. (1970). *J. Biol. Chem.* **245,** 4508–4518.

MacLennan, D. H., and Holland, P. C. (1975). *Annu. Rev. Biophys. Bioeng.* **4,** 377–404.

MacLennan, D. H., and Tzagoloff, A. (1968). *Biochemistry* **7,** 1603–1610.

MacLennan, D. H., Yip, C. C., Iles, G. H., and Seaman, P. (1972). *Cold Spring Harbor Symp. Quant. Biol.* **37,** 469–477.

MacLennan, D. H., Reithmeier, R. A. F., Shoshan, V., Campbell, K. P., and LeBel, D. (1980). *Ann. N.Y. Acad. Sci.* **358,** 138–148.

Madden, T. D., and Cullis, P. R. (1984). *J. Biol. Chem.* **259,** 7655–7658.

Madden, T. D. and Cullis, P. R. (1985). *Biochim. Biophys. Acta* **808,** 219–224.

Madden, T. D., Hope, M. J., and Cullis, P. R. (1984). *Biochemistry* **23,** 1413–1418.

Malathi, P., and Preiser, H. (1983). *Biochim. Biophys. Acta* **735,** 314–324.

Malathi, P., Preiser, H., and Crane, R. K. (1980). *Ann. N.Y. Acad. Sci.* **358,** 253–266.

Malpartida, F., and Serrano, R. (1981a). *Eur. J. Biochem.* **116,** 413–417.

Malpartida, F., and Serrano, R. (1981b). *J. Biol. Chem.* **256,** 4175–4177.

Malysheva, M. K., Lishko, V. K., and Chagovetz, A. M. (1980). *Biochim. Biophys. Acta* **602,** 70–77.

Maness, P. F., and Levy, B. T. (1983). *Mol. Cell. Biol.* **3,** 102–112.

Marchesi, V. T., and Steers, E., Jr. (1968). *Science* **159,** 203–204.

Marcus, M. A., Lewis, A., Racker, E., and Crespi, H. (1977). *Biochem. Biophys. Res. Commun.* **78,** 669–675.

Maron, R., Fishkes, H., Kanner, B. I., and Schuldiner, S. (1979). *Biochemistry* **18,** 4781–4785.

Marquardt, H., Hunkapiller, M. W., Hood, L. E., Twardzik, D. R., De Larco, J. E., Stephenson, J. R., and Todaro, G. J. (1983). *Proc. Natl. Acad. Sci. U.S.A.* **80,** 4684–4688.

Marshall, C. F., and Sager, R. (1981). *Somatic Cell Genet.* **7,** 713–723.

Matsushita, K., Patel, L., Gennis, R. B., and Kaback, H. R. (1983). *Fed. Proc., Fed. Am. Soc. Exp. Biol.* **42,** 1078 (abstr.).

Mazurek, N., Bashkin, P., Loyter, A., and Pecht, I. (1983). *Proc. Natl. Acad. Sci. U.S.A.* **80,** 6014–6018.

Mazurek, N., Schindler, H., Schürholz, T., and Pecht, I. (1984). *Proc. Natl. Acad. Sci. U.S.A.* **81,** 6841–6845.

Mego, J. L., Farb, R. M., and Barnes, J. (1972). *Biochem. J.* **128,** 763–769.

Mela, L. (1968). *Arch. Biochem. Biophys.* **123,** 286–293.

Melchior, D. L., and Czech, M. P. (1979). *J. Biol. Chem.* **254,** 8744–8747.

Metcalfe, J. C., and Warren, G. B. (1977). *In* "International Cell Biology" (B. R. Brinkley and K. R. Porter, eds.), pp. 15–23. Rockefeller Univ. Press, New York.

Metzger, H. (1983). *In* "Contemporary Topics in Molecular Immunology" (F. P. Inman and T. J. Kindt, eds.), pp. 115–145. Plenum, New York.

Meyer, D. I., and Dobberstein, B. (1980). *J. Cell Biol.* **87,** 498–502.

Meyer, D. I., Krause, E., and Dobberstein, B. (1982). *Nature (London)* **297,** 647–650.

Meyerhof, O. (1927). *Biochem. Z.* **183,** 176–215.

Meyerhof, O. (1945). *J. Biol. Chem.* **157,** 105–119.

Michel, H. (1983). *Trends Biochem. Sci.* **8,** 56–59.

Michell, R. H. (1975). *Biochim. Biophys. Acta* **415,** 81–147.

Michell, R. H., Kirk, C. J., Jones, L. M., Downes, C. R., and Creba, J. A. (1981). *Philos. Trans. R. Soc. London, Ser. B.* **296,** 123–138.

Miller, C. (1982) *In* "Transport in Biomembranes: Model Systems and Reconstitution" (R. Antolini *et al.,* eds.), pp. 99–108. Raven Press, New York.

Miller, C. (1983). *Physiol. Rev.* **63,** 1209–1242.

Miller, C., and Racker, E. (1976a). *J. Membr. Biol.* **26,** 319–333.

Miller, C., and Racker, E. (1976b). *J. Membr. Biol.* **30,** 283–300.

Miller, C., and Racker, E. (1979). *In* "The Receptors: A Comprehensive Treatise" (R. D. O'Brien, ed.), Vol. 1, pp. 1–31. Plenum, New York.

Miller, C., Arvan, P., Telford, J. N., and Racker, E. (1976). *J. Membr. Biol.* **30,** 271–282.

Miller, M. S., Benore-Parsons, M., and White, H. B., III (1982). *J. Biol. Chem.* **257,** 6818–6824.

Mills, J. D., Mitchell, P., and Schürmann, P. (1980). *FEBS Lett.* **112,** 173–177.

Mishina, M., Kurosaki, T., Tobimatsu, T., Morimoto, Y., Noda, M., Yamamoto, T., Terao, M., Lindstrom, J., Takahashi, T., Kuno, M., and Numa, S. (1984). *Nature (London)* **307,** 604–608.

Mitchell, C. D., and Hanahan, D. J. (1966). *Biochemistry* **5,** 51–57.

Mitchell, P. (1966). *Biol. Rev. Cambridge Philos. Soc.* **41,** 445–502.

Mitchell, P. (1979). *Science* **206,** 1148–1159.

Mitchell, P., and Koppenol, W. H. (1982). *Ann. N.Y. Acad. Sci.* **402,** 584–601.

Mitchell, P., and Moyle, J. (1969). *Eur. J. Biochem.* **9,** 149–155.

Miyamoto, H., and Racker, E. (1980). *J. Biol. Chem.* **255,** 2656–2658.

Miyamoto, H., and Racker, E. (1981). *FEBS Lett.* **133,** 235–238.

Miyao, M., and Murata, N. (1983). *FEBS Lett.* **164,** 375–378.

Moczydlowski, E. G., and Fortes, P. A. G. (1981). *J. Biol. Chem.* **256,** 2346–2356.

Montal, M., Darszon, A., and Schindler, H. (1981). *Q. Rev. Biophys.* **14,** 1–79.

Moolenaar, W. H., Tertoolen, L. G. J., and de Laat, S. W. (1984). *Nature (London)* **312,** 371–374.

Moore, C. L. (1971). *Biochem. Biophys. Res. Commun.* **42**, 298–305.

Morton, R. K. (1955). *In* "Methods in Enzymology" 1, pp. 25–51. (S.P. Colowick and N. O. Kaplan, eds.), Academic Press, New York.

Mosher, M. E., Peters, L. K., and Fillingame, R. H. (1983). *J Bacteriol.* **156**, 1078–1092.

Mueller, P. (1975). *In* "Energy Transducing Mechanisms" (E. Racker, ed.), Vol. 3, pp. 75–120. Butterworth, London.

Mullet, J. E., Baldwin, T. O., and Arntzen, C. J. (1981). *In* "Photosynthesis. III. Structure and Molecular Organization of the Photosynthetic Apparatus" (G. Akoyunoglou, ed.), pp. 577–582. Balaban Int. Sci. Serv., Philadelphia, Pennsylvania.

Mullins, L. J. (1976). *Fed. Proc., Fed. Am. Soc. Exp. Biol.* **35**, 2583–2588.

Mullins, R. E., and Langdon, R. G. (1980). *Biochemistry* **19**, 1199–1205.

Murata, N., Miyao, M., Omata, T., Matsunami, H., and Kuwabara, T. (1984). *Biochim. Biophys. Acta* **765**, 363–369.

Nakae, T. (1976). *J. Biol. Chem.* **251**, 2176–2178.

Nakamura, S., and Racker, E. (1984). *Biochemistry* **23**, 385–389.

Nakashima, R. A., and Garlid, K. S. (1982). *J. Biol. Chem.* **257**, 9252–9254.

Nalecz, M. J., Casey, R. P., and Azzi, A. (1983). *Biochim. Biophys. Acta* **724**, 75–82.

Navarro, J., Toivio-Kinnucan, M., and Racker, E. (1984). *Biochemistry* **23**, 130–135.

Navarro, J., Chabot, J., Sherrill, K., Aneja, R., Zahler, S. A., and Racker, E. (1985). *Biochemistry* **24**, 4645–4650.

Nedivi, E., and Schramm, M. (1984). *J. Biol. Chem.* **259**, 5803–5808.

Nelson, N. (1981a). *Curr. Top. Bioenerg.* **11**, 1–33.

Nelson, N. (1981b). *In* "Chemiosmotic Proton Circuits in Biological Membranes" (V. P. Skulachev and P. C. Hinkle, eds.), pp. 471–480. Addison-Wesley, Reading, Massachusetts.

Nelson, N., and Hauska, G. (1979). *In* "Membrane Bioenergetics" (C. P. Lee, G. Schatz, and L. Ernster, eds.), pp. 189–202. Addison-Wesley, Reading, Massachusetts.

Nelson, N., and Karny, O. (1976). *FEBS Lett.* **70**, 249–253.

Nelson, N., and Schatz, G. (1979a). *In* "Membrane Bioenergetics" (C. P. Lee, G. Schatz, and L. Ernster, eds.), pp. 133–152. Addison-Wesley, Reading, Massachusetts.

Nelson, N., and Schatz, G. (1979b). *Proc. Natl. Acad. Sci. U.S.A.* **76**, 4365–4369.

Nelson, N., Nelson, H., and Racker, E. (1972). *J. Biol. Chem.* **247**, 7657–7662.

Nelson, N., Deters, D. W., Nelson, H., and Racker, E. (1973). *J. Biol. Chem.* **248**, 2049–2055.

Nelson, N., Eytan, E., Notsani, B.-E., Sigrist, H., Sigrist-Nelson, K., and Gilter, C. (1977). *Proc. Natl. Acad. Sci. U.S.A.* **74**, 2375–2378.

Nelson, S. O., and Postma, P. W. (1984). *Eur. J. Biochem.* **139**, 29–34.

Nelson, S. O., Wright, J. K., and Postma, P. W. (1983). *EMBO J.* **2**, 715–720.

Neubig, R., Krodel, E. K., Boyd, N. D., and Cohen, J. B. (1979). *Proc. Natl. Acad. Sci. U.S.A.* **76**, 690–694.

Neufeld, E. F. (1981). *In* "Lysosomes and Lysosomal Storage Diseases" (J. W. Callahan and J. A. Lowden, eds.), pp. 115–129. Raven Press, New York.

Neufeld, E. F., Lim, T. W., and Shapiro, L. J. (1975). *Annu. Rev. Biochem.* **44**, 357–376.

Neufeld, G., Schramm, M., and Weinberg, N. (1980). *J. Biol. Chem.* **255**, 9268–9274.

Neupert, W., and Schatz, G. (1981). *Trends Biochem. Sci.* **6,** 1–4.

Newman, M. J., and Wilson, T. H. (1980). *J. Biol. Chem.* **255,** 10583–10586.

Newman, M. J., Foster, D. L., Wilson, T. H., and Kaback, H. R. (1981). *J. Biol. Chem.* **256,** 11804–11808.

Nexo, E., Hock, R. A., and Hollenberg, M. D. (1979). *J. Biol. Chem.* **254,** 8740–8743.

Nicholls, D. G. (1979). *Biochim. Biophys. Acta* **549,** 1–29.

Nicholls, D. G. (1983). *Biosci. Rep.* **3,** 431–441.

Nicholls, D. G., and Akerman, K. (1982). *Biochim. Biophys. Acta* **683,** 57–88.

Nicholls, P. (1981). *Int. Rev. Cytol., Suppl.* **12,** 327–388.

Nielson, J., Hansen, F. G., Hoppe, J., Friedl, P., and Meyenburg, K. (1981). *Mol. Gen. Genet.* **184,** 33–39.

Niggli, V., Sigel, E., and Carafoli, E. (1982). *J. Biol. Chem.* **257,** 2350–2356.

Nikaido, H., and Nakae, T. (1979). *Adv. Microb. Physiol.* **20,** 163–250.

Nilsen-Hamilton, M., and Hamilton, R. T. (1979). *Biochim. Biophys. Acta* **588,** 322–331.

Nishibayashi-Yamashita, H., Cunningham, C., and Racker, E. (1972). *J. Biol. Chem.* **247,** 698–704.

Nishino, H., Tillotson, L. G., Schiller, R. M., Inui, K-I., and Isselbacher, K. J. (1980). *Arch. Biochem. Biophys.* **203,** 428–436.

Nishizuka, Y. (1984). *Trends Biochem. Sci.* **9,** 163–166.

Nozaki, Y., and Tanford, C. (1981). *Proc. Natl. Acad. Sci. U.S.A.* **78,** 4324–4328.

Nuccitelli, R., and Deamer, D. W., eds. (1982). "Intracellular pH: Its Measurement Regulation and Utilization in Cellular Functions." Alan R. Liss, Inc., New York.

Ochoa, E. L. M., Dalziel, A. W., and McNamee, M. G. (1983). *Biochim. Biophys. Acta* **727,** 151–162.

O'Donnell-Tormey, J., and Quigley, J. P. (1983). *Proc. Natl. Acad. Sci. U.S.A.* **80,** 344–348.

Oesterhelt, D., and Christoffel, V. (1976). *Biochem. Soc. Trans.* **4,** 556–559.

Oesterhelt, D., and Stoeckenius, W. (1973). *Proc. Natl. Acad. Sci. U.S.A.* **70,** 2853–2857.

Ohkuma, S., Moriyama, Y., and Takano, T. (1982). *Proc. Natl. Acad. Sci. U.S.A.* **79,** 2758–2762.

Ohnishi, T. (1979). *In* "Membrane Proteins in Energy Transductions" (R. A. Capaldi, ed.), pp. 1–87. Dekker, New York.

Okamota, H., Sone, N., Hirata, H., Yoshida, M., and Kagawa, Y. (1977). *J. Biol. Chem.* **252,** 6125–6131.

Oku, N., and MacDonald, R. C. (1983). *Biochemistry* **22,** 855–863.

Oleszko, S., and Moudrianakis, E. N. (1974). *J. Cell Biol.* **63,** 936–948.

O'Neal, S. G., Rhoads, D. B., and Racker, E. (1979). *Biochem. Biophys. Res. Commun.* **89,** 845–850.

O'Neill, S. D., and Spanswick, R. M. (1984). *J. Membr. Biol.* **79,** 231–243.

Orlich, G., and Hauska, G. (1980). *Eur. J. Biochem.* **111,** 525–533.

Ossowski, L., and Reich, E. (1983). *Cell* **35,** 611–619.

Ostro, M. J. (1983). *In* "Liposome letters" (A. D. Bangham, ed.) pp. 309–317. Academic Press, New York.

Otsuka, M., Ohtsuki, I., and Ebashi, S. (1965). *J. Biochem. (Tokyo)* **58,** 188–190.

Ovchinnikov, Y. A., Abdulaev, N. G., Feigina, M. Y. A., Kiselev, A. V., and Lobanov, N. A. (1979). *FEBS Lett.* **100,** 219–224.

Ozanne, B., Fulton, R. J., and Kaplan, P. L. (1980). *J. Cell. Physiol.* **105,** 163–180.

Pachence, J. M., Dutton, P. L., and Blasie, J. K. (1983). *Biochim. Biophys. Acta* **724,** 6–19.

Panfili, E., Sottocasa, G. L., Sandri, G., and Luit, G. (1980). *Eur. J. Biochem.* **105,** 205–210.

Pappenheimer, A. M., Jr. (1977). *Annu. Rev. Biochem.* **46,** 69–94.

Pastan, I., and Willingham, M. C. (1981). *Annu. Rev. Physiol.* **43,** 239–250.

Pearse, B. M. F., and Bretscher, M. S. (1981). *Annu. Rev. Biochem.* **50,** 85–101.

Pedersen, P. L. (1982). *Ann. N. Y. Acad. Sci.* **402,** 1–20.

Peerce, B. E., and Wright, E. M. (1984). *Proc. Natl. Acad. Sci. U.S.A.* **81,** 2223–2226.

Penefsky, H. S. (1979). *Adv. Enzymol.* **49,** 223–280.

Penefsky, H. S., and Tzagoloff, A. (1971). *In* "Methods in Enzymology" (W. B. Jakoby, ed.), Vol. 22, pp.204–218. Academic Press, New York.

Pennington, R. M., and Fisher, R. R. (1981). *J. Biol. Chem.* **256,** 8963–8969.

Penniston, J. T. (1982). *Ann. N. Y. Acad. Sci.* **402,** 296–303.

Perlin, D. S., Cox, D. N., and Senior, A. E. (1983). *J. Biol. Chem.* **258,** 9793–9800.

Peters, W. H. M., Fleuren-Jakobs, A. M. M., Schrijen, J. J., De Pont, J. J. H. H. M., and Bonting, S. L. (1982). *Biochim. Biophys. Acta* **690,** 251–260.

Peterson, G. L., Churchill, L., Fisher, J. A., and Hokin, L. E. (1982). *Ann. N. Y. Acad. Sci.* **402,** 185–206.

Phelps, D. C., and Hatefi, Y. (1981). *J. Biol. Chem.* **256,** 8217–8221.

Philosoph, S., Khananshvili, D., and Gromet-Elhanan, Z. (1981). *Biochem. Biophys. Res. Commun.* **101,** 384–389.

Pick, U. (1981). *Arch. Biochem. Biophys.* **212,** 186–194.

Pick, U., and Racker, E. (1979a). *J. Biol. Chem.* **254,** 2793–2799.

Pick, U., and Racker, E. (1979b). *Biochemistry* **18,** 108–113.

Pickart, C. M., and Jencks, W. P. (1984). *J. Biol. Chem.* **259,** 1629–1643.

Pitotti, A., Dabbeni-Sala, F., and Bruni, A. (1980). *Biochim. Biophys. Acta* **600,** 79–90.

Pitts, B. J. R. (1979). *J. Biol. Chem.* **254,** 6232–6235.

Plagemann, P. G. W., and Wohlhueter, R. M. (1980). *Curr. Top. Membr. Transp.* **14,** 225–330.

Pledger, W. J., Stiles, C. D., Antoniades, H. N., and Scher, C. D. (1978). *Proc. Natl. Acad. Sci. U.S.A.* **75,** 2839–2843.

Poole, R. J. (1978). *Annu. Rev. Plant Physiol.* **29,** 437–460.

Popot, J.-L. (1983). *In* "Basic Mechanisms of Neuronal Hyperexcitability," pp. 137–170. Alan R. Liss, Inc., New York.

Popot, J.-L., Cartaud, J., and Changeux, J. P. (1981). *Eur. J. Biochem.* **118,** 203–214.

Portis, A. R., Jr., and McCarty, R. E. (1974). *J. Biol. Chem.* **249,** 6250–6254.

Poste, G. (1983). *In* "Liposomes Letters" (A. D. Bangham, ed.), pp. 375–386. Academic Press, New York.

Postma, P. W., and Roseman, S. (1976). *Biochim. Biophys. Acta* **457,** 213–257.

Pouysségur, J., Franchi, A., Salomon, J.-C., and Silvestre, P. (1980). *Proc. Natl. Acad. Sci. U.S.A.* **77,** 2698–2701.

Pressman, B. C. (1976). *Annu. Rev. Biochem.* **45,** 501–530.

Prince, R. C., Matsuura, K., Hurt, E., Hauska, G., and Dutton, P. L. (1982). *J. Biol. Chem.* **257,** 3379–3381.

Puettner, I., Carafoli E., and Malatesta, F. (1985). *J. Biol. Chem.* **260,** 3719–3723.

Pullman, M. E., and Monroy, G. C. (1963). *J. Biol. Chem.* **238,** 3762–3769.

Pullman, M. E., Penefsky, H. S., Datta, A., and Racker, E. (1960). *J. Biol. Chem.* **235**, 3322–3329.

Puskin, J. S., Gunter, T. E., Gunter, K. K., and Russell, P. R. (1976). *Biochemistry* **15**, 3834–3842.

Quinton, P. M. (1983). *Nature* **301**, 421–422.

Racker, E. (1955). *Nature (London)* **175**, 249–255.

Racker, E. (1962). *Proc. Natl. Acad. Sci. U.S.A.* **48**, 1659–1663.

Racker, E. (1963). *Biochem. Biophys. Res. Commun.* **10**. 435–439.

Racker, E. (1965). "Mechanisms in Bioenergetics." Academic Press, New York.

Racker, E. (1970). *In* "Membranes of Mitochondria and Chloroplasts" (E. Racker, ed.), pp. 127–171. Van Nostrand-Reinhold, Princeton, New Jersey.

Racker, E. (1972a). *J. Membr. Biol.* **10**, 221–235.

Racker, E. (1972b). *J. Biol. Chem.* **247**, 8198–8200.

Racker, E. (1973). *Biochem. Biophys. Res. Commun.* **55**, 224–230.

Racker, E. (1976). "A New Look at Mechanisms in Bioenergetics." Academic Press, New York.

Racker, E. (1977a). *Annu. Rev. Biochem.* **46**, 1006–1014.

Racker, E. (1977b). *In* "Calcium Binding Proteins and Calcium Function" (R. H. Wasserman *et al.,* eds.), pp. 155–163. Elsevier/North-Holland Biomedical Press, Amsterdam.

Racker, E. (1978). *In* "Membrane Transport in Biology" (G. Giebisch, D. C. Tosteson, and H. H. Ussing, eds.), Vol. I, pp. 259–290. Springer-Verlag, Berlin and New York.

Racker, E. (1979). *In* "Methods in Enzymology" (S. Fleischer and L. Packer, eds.), Vol. 55, pp. 699–711. Academic Press, New York.

Racker, E. (1980). *Fed. Proc., Fed. Am. Soc. Exp. Biol.* **39**, 2422–2426.

Racker, E. (1981a). *In* "Mitochondria and Microsomes" (C. P. Lee, G. Schatz, and G. Dallner, eds.), pp. 337–355. Addison-Wesley, Reading, Massachusetts.

Racker, E. (1981b). *Science* **213**, 1313 (Retraction Letter).

Racker, E. (1981c). *Curr. Top. Cell. Regul.* **18**, 361–376.

Racker, E. (1983a). *Fed. Proc., Fed. Am. Soc. Exp. Biol.* **42**, 2899–2909.

Racker, E. (1983b). *Biosci. Rep.* **3**, 507–516.

Racker, E. (1983c). *In* "Liposome Letters" (A. D. Bangham, ed.), pp. 179–195. Academic Press, New York.

Racker, E. (1983d). *Science* **222**, 232.

Racker, E. (1985). *In* "Frontiers of Membrane Research in Agriculture" BARC IX, (J. St. John, ed.), in press.

Racker, E. and Eytan, E. (1973). *Biochem. Biophys. Res. Commun.* **55**, 174–158.

Racker, E., and Eytan, E. (1975). *J. Biol. Chem.* **250**, 7533–7534.

Racker, E., and Fisher, L. W. (1975). *Biochem. Biophys. Res. Commun.* **67**, 1144–1150.

Racker, E., and Hinkle, P. C. (1974). *J. Membr. Biol.* **17**, 181–188.

Racker, E., and Horstman, L. L. (1967). *J. Biol. Chem.* **242**, 2547–2551.

Racker, E., and Horstman, L. L. (1972). *In* "Energy Metabolism and the Regulation of Metabolic Processes in Mitochondria" (M. A. Mehlman and R. W. Hanson, eds.), pp. 1–25. Academic Press, New York.

Racker, E., and Kandrach, A. (1971). *J. Biol. Chem.* **246**, 7069–7071.

Racker, E., and Kandrach, A. (1973). *J. Biol. Chem.* **248**, 5841–5847.

Racker, E., and Krimsky, I. (1945). *J. Biol. Chem.* **161**, 453–461.

Racker, E., and Riegler C., (1985). *Arch. Biochem. Biophys.* **240,** 836–842.

Racker, E., and Stoeckenius, W. (1974). *J. Biol. Chem.* **249,** 662–663.

Racker, E., Klybas, V., and Schramm, M. (1959). *J. Biol. Chem.* **234,** 2510–2516.

Racker, E., Horstman, L. L., Kling, D., and Fessenden-Raden, J. M. (1969). *J. Biol. Chem.* **244,** 6668–6674.

Racker, E., Fessenden-Raden, J. M., Kandrach, M. A., Lam, K. W., and Sanadi, D. R. (1970). *Biochem. Biophys. Res. Commun.* **41,** 1474–1479.

Racker, E., Chien, T.-F., and Kandrach, A. (1975). *FEBS Lett.* **57,** 14–18.

Racker, E., Violand, B., O'Neal, S., Alfonzo, M., and Telford, J. (1979). *Arch. Biochem. Biophys.* **198,** 470–477.

Racker, E., Belt, J. A., Carley, W. W., and Johnson, J. H. (1980a). *Ann. N. Y. Acad. Sci.* **341,** 27–36.

Racker, E., Miyamoto, H., Mogerman, J., Simons, J., and O'Neal, S. (1980b). *Ann. N. Y. Acad. Sci.* **358,** 64–72.

Racker, E., Johnson, J. H., and Thiery Blackwell, M. (1983). *J. Biol. Chem.* **258,** 3702–3705.

Racker, E., Riegler, C., and Abdel-Ghany, M. (1984). *Cancer Res.* **44,** 1364–1367.

Racker, E., Resnick, R. J., and Feldman, R. (1985). *Proc. Natl. Acad. Sci. U.S.A.* **82,** 3535–3538.

Ragan, C. I., and Hinkle, P. C. (1975) *J. Biol. Chem.* **250,** 8472–8476.

Ragan, C. I., and Racker, É. (1973). *J. Biol. Chem.* **248,** 6876–6884.

Ragan, C. I., and Widger, W. R. (1975). *Biochem. Biophys. Res. Commun.* **62,** 744–749.

Ragan, C. I., Smith, S., Earley, F. G. P., and Poore, V. M. (1981). *In* "Chemiosmotic Proton Circuits in Biological Membranes" (V. P. Skulachev and P. C. Hinkle, eds.), pp. 59–68. Addison-Wesley, Reading, Massachusetts.

Reeves, A. S., Collins, J. H., and Schwartz, A. (1980). *Biochem. Biophys. Res. Commun.* **95,** 1591–1598.

Reeves, J., and Dowben, R. (1969). *J. Cell. Physiol.* **73,** 49.

Reeves, J. P., and Sutko, J. L. (1979). *Proc. Natl. Acad. Sci. U.S.A.* **76,** 590–594.

Rehm, W. S. (1965). *Fed. Proc., Fed. Am. Soc. Exp. Biol.* **24,** 1387–1395.

Resh, M. D., Nemenoff, R. A., and Guidotti, G. (1980). *J. Biol. Chem.* **255,** 10938–10945.

Reynolds, J. A., and Karlin, A. (1978). *Biochemistry* **17,** 2035–2038.

Reynolds, J. A., and Trayer, H. (1971). *J. Biol. Chem.* **246,** 7337–7342.

Rhee, H. M., and Hokin, L. E. (1975). *Biochem. Biophys. Res. Commun.* **63,** 1139–1145.

Rhoads, D. B., and Epstein, W. (1977). *J. Biol. Chem.* **252,** 1394–1401.

Rhodin, T. R., and Racker, E. (1974). *Biochem. Biophys. Res. Commun.* **61,** 1207–1212.

Rice, W. R., and Steck, T. L. (1976). *Biochim. Biophys. Acta* **433,** 39–53.

Rich, P. R., and Heathcote, P. (1983). *Biochim. Biophys. Acta* **725,** 332–340.

Richter, M. L., Patrie, W. J., and McCarty, R. E. (1984). *J. Biol. Chem.* **259,** 7371–7373.

Riezman, H., Hay, R., Witte, C., Nelson, N., and Schatz, G. (1983). *EMBO J.* **2,** 1113–1118.

Rigaud, J. L., Bluzat, A., and Buschlen, S. (1983). *Biochem. Biophys. Res. Commun.* **111,** 373–382.

Riklis, E., and Quastel, J. H. (1958). *Can. J. Biochem. Physiol.* **36,** 347.

Rindler, M. J., and Saier, M. H., Jr. (1981). *J. Biol. Chem.* **256,** 10820–10825.

Roberts, A. B., Frolik, C. A., Anzano, M. A., and Sporn, M. B. (1983). *Fed. Proc.,* *Fed. Am. Soc. Exp. Biol.* **42,** 2621–2626.

Roberts, A. B., Anzano, M. A., Assoian, R. K., Frolik, C. A., and Sporn, M. B. (1984). *Proc. Am. Assoc. Cancer Res.* **25,** 414–415.

Robertson, D. E., Kaczorowski, G. J., Garcia, M.-L., and Kaback, H. R. (1980). *Biochemistry* **19,** 5692–5702.

Rögner, M., Ohno, K., Hamamoto, T., Sone, N., and Kagawa, Y. (1979). *Biochem. Biophys. Res. Commun.* **91,** 362–367.

Roisin, M. P., Scherman, D., and Henry, J. P. (1980). *FEBS Lett.* **115,** 143–147.

Roos, A., and Boron, W. F. (1981). *Physiol. Rev.* **61,** 296–434.

Roseman, S. *et al.* (1982). *J. Biol. Chem.* **257,** 14461–14575.

Rosen, O. M., Herrera, R., Olowe, Y., Petruzzelli, L. M., and Cobb, M. H. (1983). *Proc. Natl. Acad. Sci. U.S.A.* **80,** 3237–3240.

Rosenberg, R. L., Tomiko, S. A., and Agnew, W. S. (1984). *Proc. Natl. Acad. Sci. U.S.A.* **81,** 1239–1243.

Rosenbusch, J. P. (1974). *J. Biol. Chem.* **249,** 8019–8029.

Rothman, J. E. (1981). *Science* **213,** 1212–1219.

Rothman, J. E., and Leonard, J. (1984). *Trends Biochem. Sci.* **9,** 176–178.

Rottenberg, H. (1983). *Proc. Natl. Acad. Sci. U.S.A.* **80,** 3313–3317.

Rozengurt, E. (1983). *Mol. Biol. Med.* **1,** 169–181.

Rozengurt, E. (1984). *Horm. Cell Regul.* **8,** 17–36.

Rozengurt, E., and Heppel, L. A. (1975). *Proc. Natl. Acad. Sci. U.S.A.* **72,** 4492–4495.

Rŭdnick, G., and Nelson, P. J. (1978). *Biochemistry* **17,** 5300–5303.

Rüegg, U. T., Hiller, J. M., and Simon, E. J. (1980). *Eur. J. Pharmacol.* **64,** 367–368.

Rydström, J. (1977). *Biochim. Biophys. Acta* **463,** 155–184.

Rydström, J. (1979). *J. Biol. Chem.* **254,** 8611–8619.

Rydström, J., Kanner, N., and Racker, E. (1975). *Biochem. Biophys. Res. Commun.* **67,** 831–839.

Rydström, J., Lee, C. P., and Ernster, L. (1981). *In* "Chemiosmotic Proton Circuits in Biological Membranes" (V. P. Skulachev and P. C. Hinkle, eds.), pp. 483–508. Addison-Wesley, Reading, Massachusetts.

Saccomani, G., Dailey, D. W., and Sachs, G. (1979). *J. Biol. Chem.* **254,** 2821–2827.

Saccomani, G., Sachs, G., Cupolletti, J., and Jung, C. Y. (1981). *J. Biol. Chem.* **256,** 7727–7729.

Sachs, G., Rabbon, E., and Saccomani, G. (1979). *In* "Cation Flux across Biomembranes" (Y. Mukohata and L. Packer, eds.), pp. 53–66. Academic Press, New York.

Sacktor, B. (1982). *In* "Membranes and Transport" (A. N. Martonosi, ed.), Vol. 2, pp. 197–206. Plenum, New York.

Sager, R. (1982). *In* "Tumor Cell Heterogeneity" (A. H. Owens, Jr., D. S. Coffey, and S. B. Baylin, eds), pp. 411–423. Academic Press, New York.

Saier, M. H., Jr. (1982). *In* "Membranes and Transport" (A. N. Martonosi, ed.), Vol. 2, pp. 27–32. Plenum, New York.

Sakamoto, J., and Tonomura, Y. (1983). *J. Biochem. (Tokyo)* **93,** 1601–1614.

Salmon, S. E., and von Hoff, D. D. (1981). *Semin. Oncol.* **8,** 377–385.

Salter, D. W., Baldwin, S. A., Lienhard, G. E., and Weber, M. J. (1982). *Proc. Natl. Acad. Sci. U.S.A.* **79,** 1540–1544.

Saltzgaber-Müller, J., Douglas, M., and Racker, E. (1980). *FEBS Lett.* **120**, 49–52.
Sanadi, D. R., Lam, K. W., and Ramakrishna Kurup, C. K. (1968). *Proc. Natl. Acad. Sci. U.S.A.* **61**, 277–283.
Sandvig, K., and Olsnes, S. (1984). *J. Cell. Physiol.* **119**, 7–14.
San Pietro, A., and Lang, H. M. (1958). *J. Biol. Chem.* **231**, 211–229.
Saraste, M., Penttilä, T., and Wikström, M. (1981). *Eur. J. Biochem.* **115**, 261–268.
Sariban-Sohraby, S., Latorre, R., Burg, M., Olans, L., and Benos, D. (1984). *Nature (London)* **308**, 80–82.
Scarborough, G. A. (1978). *Methods Cell Biol.* **20**, 117–133.
Schatz, G., and Mason, T. L. (1974). *Annu Rev. Biochem.* **43**, 51–87.
Schatz, G., Penefsky, H. S., and Racker, E. (1967). *J. Biol. Chem.* **242**, 2552–2560.
Schatzmann, H. J. (1982). *In* "Membrane Transport of Calcium" (E. Carafoli, ed.), pp. 41–108. Academic Press, London.
Schekman, R., Esmon, B., Ferro-Novick, S., Field, C., and Novick, P. (1983). *In* "Methods in Enzymology" (S. Fleischer and B. Fleischer, eds.), Vol. 96, pp. 802–815.Academic Press, New York.
Schenerman, M. A., Racker, E., and Kilberg, M. S. (1985). *J. Cell Biol.*, in press.
Schindler, H., and Rosenbusch, J. P. (1981). *Proc. Natl. Acad. Sci. U.S.A.* **78**, 2302–2306.
Schmidt, G., and Gräber, P. (1985). *Biochim. Biophys. Acta* **808**, 46–51.
Schmidt, G., and Thannhauser, S. J. (1943). *J. Biol. Chem.* 149, 369–385.
Schneider, D. L. (1981). *J. Biol. Chem.* **256**, 3858–3864.
Schneider, D. L. (1983). *J. Biol. Chem.* **258**, 1833–1838.
Schneider, D. L., Kagawa, Y., and Racker, E. (1972). *J. Biol. Chem.* **247**, 4074–4079.
Schneider, E., and Altendorf, K. (1982). *Eur. J. Biochem.* **126**, 149–153.
Schneider, E., and Altendorf, K. (1984). *Proc. Natl. Acad. Sci. U.S.A.* **81,** 7279–7283.
Schneider, H., Fiechter, A., and Fuhrmann, G. F. (1978). *Biochim. Biophys. Acta* **512**, 495–507.
Schneider, H., Lemasters, J. J., Höchli, M., and Hackenbrock, C. R. (1980). *J. Biol. Chem.* **255**, 3748–3756.
Schobert, B., and Lanyi, J. K. (1982). *J. Biol. Chem.* **257**, 103065–10313.
Scholnick, P., Lang, D., and Racker, E. (1973) *J. Biol. Chem.* **248**, 5175–5182.
Schramm, M. (1979a). *In* "Membrane Bioenergetics" (C. P. Lee, G. Schatz, and L. Ernster, eds.), pp. 349–350. Addison-Wesley, Reading Massachusetts.
Schramm, M. (1979b). *Proc. Natl. Acad. Sci. U.S.A.* **76**, 1174–1178.
Schramm, M., Orly, J., Eimerl, S., and Korner, M. (1977). *Nature (London)* **268**, 310–313.
Schreiber, A. B., Yarden, Y., and Schlessinger, J. (1981). *Biochem. Biophys. Res. Commun.* **101**, 517–523.
Schreiber, A. B., Libermann, T. A., Lax, I., Yarden, Y., and Schlessinger, J. (1983). *J. Biol. Chem.* **258**, 846–853.
Schulz, H., and Racker, E. (1979). *Biochem. Biophys. Res. Commun.* **89**, 134–140.
Scolnick, E. M., Papageorge, A. G., and Shih, T. Y. (1979). *Proc. Natl. Acad. Sci. U.S.A.* **76**, 5355–5359.
Scott, T. L., and Shamoo, A. D. (1982). *J. Membr. Biol.* **64**, 137–144.
Sebald, W. (1977). *Biochim. Biophys. Acta* **463**, 1–27.
Sebald, W., and Hoppe, J. (1981). *Curr. Top. Bioenerg.* **12**, 1–64.
Seelig, J., Tamm, L., Hymel, L., and Fleischer, S. (1981). *Biochemistry* **20**, 3922–3932.

Senger, D. R., Asch, B. B., Smith, B. D., Perruzzi, C. A., and Dvorak, H. F. (1983). *Nature (London)* **302**, 714–715.

Senior, A. E. (1973). *Biochim. Biophys. Acta* **301**, 249–277.

Senior, A. E. (1985). *Curr. Top. Memb. Transport* **23**, 135–151.

Senior, A. E., and Wise J. G. (1983). *J. Membr. Biol.* **73**, 105–124.

Serrano, R. (1984). *Biochem. Biophys. Res. Commun.* **121**, 735–740.

Shanahan, M. F., and Czech, M. P. (1977). *J. Biol. Chem.* **252**, 8341–8343.

Shelton, R. L., Jr., and Langdon, R. G. (1983). *Biochim. Biophys. Acta* **733**, 25–33.

Shen, D.-W., Real, F. X., DeLeo, A. B., Old, L. J., Marks, P. A., and Rifkind, R. A. (1983). *Proc. Natl. Acad. Sci. U.S.A.* **80**, 5919–5922.

Sherr, C. J., Rettenmier, C. W., Sacca, R., Roussel, M. F., Look, A. T., and Stanley, E. R. (1985). *Cell* **41**, 665–676.

Shertzer, H. G., and Racker, E. (1974). *J. Biol. Chem.* **249**, 1320–1321.

Shertzer, H. G., and Racker, E. (1976). *J. Biol. Chem.* **251**, 2446–2452.

Shertzer, H. G., Kanner, B. I., Banerjee, R. K., and Racker, E. (1977). *Biochem. Biophys. Res. Commun.* **75**, 779–784.

Shin, S.-I., Freedman, V. H., Risser, R., and Pollack, R. (1975). *Proc. Natl. Acad. Sci. U.S.A.* **72**, 4435–4439.

Shorr, R. G. L., Stohsacker, M. W., Lavin, T. N., Lefkowitz, R. J., and Caron, M. G. (1982). *J. Biol. Chem.* **257**, 12341–12350.

Sidhu, A., and Beattie, D. S. (1982). *J. Biol. Chem.* **257**, 7879–7886.

Sigel, E., and Carafoli, E. (1980). *Eur. J. Biochem.* **111**, 299–306.

Sigrist-Nelson, K., and Azzi, A. (1980). *J. Biol. Chem.* **255**, 10638–10643.

Simonds, W. F., Koski, G., Streaty, R. A., Hjelmeland, L. M., and Klee, W. A. (1980). *Proc. Natl. Acad. Sci. U.S.A.* **77**, 4623–4627.

Sinozawa, T., Sen, I., Wheeler, G. L., and Bitensky, M. W. (1979). *J. Supramol. Struct.* **10**, 185–190.

Skou, J. C. (1982). *Ann. N. Y. Acad. Sci.* **402**, 169–184.

Skou, J. C., and Norby, J. G., eds. (1979). "Na, K-ATPase Structure and Kinetics." Academic Press, New York.

Skulachev, V. P. (1974). *Ann. N. Y. Acad. Sci.* **227**, 188–202.

Skulachev. V. P. (1979). *In* "Cation Flux across Biomembranes" (Y. Mukohata and L. Packer, eds.), pp. 303–319. Academic Press, New York.

Slater, E. C. (1983). *Trends Biochem. Sci.* **8**, 239–242.

Slaughter, R. S., Fenwick, R. G., Jr., and Barnes, E. M., Jr. (1981). *Arch. Biochem. Biophys.* **211**, 494–499.

Slayman, C. W. (1978). *In* "Membrane Transport in Biology" (G. Giebisch, D. C. Tosteson, and H. H. Ussing, eds.), Vol. 1, pp. 237–257. Springer-Verlag, Berlin and New York.

Sly, W. S., Natowicz, M., Gonsales-Noriega, A., Gruble, J. H., and Fischer, H. D. (1981). *In* "Lysosomes and Lysosomal Storage Diseases" (J. W. Callahan and J. A. Lowden, eds.), pp. 131–146. Raven Press, New York.

Smallwood, J. I., Waisman, D. M., Lafreniere, D., and Rasmussen, H. (1983). *J. Biol. Chem.* **258**, 11092–11097.

Smigel, M. D., Ross, E. M., and Gilman, A. G. (1984). *In* "Cell Membrane: Methods in Reviews" (E. L. Elson, W. A. Frazier, and L. Glaser, eds.), pp. 247–294. Plenum, New York.

Smith, J. B., and Sternweis, P. C. (1977). *Biochemistry* **16**, 306–311.

Smith, R., and Scarborough, G. A. (1984). *Anal. Biochem.* **138**, 156–163.

Sobel, A., Weber, M., and Changeux, J.-P. (1977). *Eur. J. Biochem.* **80,** 215–224.

Sogin, D. C., and Hinkle, P. C. (1980). *Proc. Natl. Acad. Sci. U.S.A.* **77,** 5725–5729.

Solioz, M., Carafoli, E., and Ludwig, B. (1982). *J. Biol. Chem.* **257,** 1579–1582.

Sone, N. (1982). *In* "Chemiosmotic Proton Circuits in Biological Membranes" (V. P. Skulachev and P. C. Hinkle, eds.), pp. 197–209. Addison-Wesley, Reading, Massachusetts.

Sone, N., and Yanagita, Y. (1982). *Biochim. Biophys. Acta* **682.** 216–226

Sone, N., Yoshida, M., Hirata, H., and Kagawa, Y. (1975). *J. Biol. Chem.* **250,** 7917–7923.

Sone, N., Yoshida, M., Hirata, H., and Kagawa, Y. (1978). *Proc. Natl. Acad. Sci. U.S.A.* **75,** 4219–4223.

Sone, N., Yanagita, Y., Hon-Nami, K., Fukumore, Y., and Yamanaka, T. (1983). *FEBS Lett.* **155,** 150–154.

Soumarmon, A., and Racker, E. (1978). *Front. Biol. Energ.* **1,** 555–562.

Spanswick, R. M. (1981). *Annu. Rev. Plant Physiol.* **32,** 267–289.

Spatz, L., and Strittmatter, P. (1971). *Proc. Natl. Acad. Sci. U.S.A.* **68,** 1042–1046.

Spector, M., O'Neal, S., and Racker, E. (1980). *J. Biol. Chem.* **255,** 8370–8373.

Spencer, T. L., and Lehninger, A. L. (1976). *Biochem. J.* **154,** 405–415.

Sporn, M. B., and Todaro, G. J. (1980). *N. Engl. J. Med.* **303,** 878–880.

Stanley, W. M., and Knight, C. A. (1941). *Cold Spring Harbor Symp. Quant. Biol.* **9,** 255–262.

Stern, P. H., Wallace, C. D., and Hoffman, R. M. (1984). *J. Cell. Physiol.* **119,** 29–34.

Sternweis, P. C., Northrup, J. K., Smigel, M. D., and Gilman, A. G. (1981). *J. Biol. Chem.* **256,** 11517–11526.

Stiles, G. L., Strasser, R. H., Caron, M. G., and Lefkowitz, R. J. (1983). *J. Biol. Chem.* **258,** 10689–10694.

Stipani, I., and Palmieri, F. (1983). *FEBS Lett.* **161,** 269–274.

Stipani, I., Zara, V., Iacobazzi, V., and Palmieri, F. (1984). *Ital. J. Biochem.* **33,** 218–220.

Stone, D. K., Xie, X.-S., and Racker, E. (1983a). *J. Biol. Chem.* **258,** 4059–4062.

Stone, D. K., Seldin, D. W., Kokko, J. P., and Jacobson, H. R. (1983b). *J. Clin. Invest.* **72,** 77–83.

Stone, D. K., Xie, X.-S., Wu, L.-T., and Racker, E. (1984a). *In* "Hydrogen Ion Transport in Epithelia" (J. G. Forte, D. G. Warnock, and F. C. Rector, Jr., eds.), pp. 219–230. Wiley, New York.

Stone, D. K., Xie, X.-S., and Racker, E. (1984b). *J. Biol. Chem.* **259,** 2701–2703.

Strapazon, E., and Steck, T. L. (1976). *Biochemistry* **15,** 1421–1424.

Strosberg, A. D. (1984). *Trends Biochem. Sci.* **9,** 166–169.

Stryer, L., Hurley, J. B., and Fung, B. K.-K. (1981). *Trends Biochem. Sci.* **6,** 245–247.

Šubík, J., Kolarov, J., and Kováč, L. (1974). *Biochim. Biophys. Acta* **357,** 453–456.

Sugimoto, Y., Whitman, M., Cantley, L. C., and Erikson, R. L. (1984). *Proc. Natl. Acad. Sci. U.S.A.* **81,** 2117–2121.

Suhadolnik, R. J. (1979). "Nucleosides as Biological Probes." Wiley, New York.

Suzuki, K., and Kono, T. (1980). *Proc. Natl. Acad. Sci. U.S.A.* **77,** 2542–2545.

Sweadner, K. T. (1979). *J. Biol. Chem.* **254,** 6060–6067.

Sweadner, K. T., and Goldin, S. M. (1975). *J. Biol. Chem.* **250,** 4022–4024.

Szoka, F., Jr., and Papahadjopoulos, D. (1978). *Proc. Natl. Acad. Sci. U.S.A.* **75,** 4194–4198.

Tada, M., Kirchberger, M. A., Repke, D. I., and Katz, A. M. (1974). *J. Biol. Chem.* **249**, 6174–6180.

Takeda, K., Hirano, M., Kanazawa, H., Nukiwa, N., Kagawa, Y., and Futai, M. (1982). *J. Biochem. (Tokyo)* **91**, 695–701.

Talvenheimo, J., and Rŭdnick, G. (1980). *J. Biol. Chem.* **255**, 8606–8611.

Talvenheimo, J., Fishkes, H., Nelson, P. J., and Rŭdnick, G. (1983). *J. Biol. Chem.* **258**, 6115–6119.

Tamkun, M. M., Talvenheimo, J. A., and Catterall, W. A. (1984). *J. Biol. Chem.* **259**, 1676–1688.

Tanford, C., and Reynolds, J. A. (1976). *Biochim. Biophys. Acta* **457**, 133–170.

Tank, D. W., Miller, C., and Webb, W. W. (1982). *Proc. Natl. Acad. Sci. U.S.A.* **79**, 7749–7753.

Taub, M. (1978). *Somatic Cell Gent.* **4**, 609–616.

Taylor, W. M., Prpíc, V., Exton, J. H., and Bygrave, F. L. (1980). *Biochem. J.* **188**, 443–450.

Teather, R. M., Bramhall, J., Riede, I., Wright, J. K., Fürst, M., Aichele, G., Wilhelm, U., and Overath, P. (1980). *Eur. J. Biochem.* **108**, 223–231.

Teintze, M., Slaughter, M., Weiss, H., and Neupert, W. (1982). *J. Biol. Chem.* **257**, 10364–10371.

Telford, J. N., and Racker, E. (1973). *J. Cell Biol.* **57**, 580–586.

Telford, J. N., Langworthy, T. A., and Racker, E. (1984). *J. Bioenerg. Biomembr.* **16**, 335–351.

Temin, H. M., Pierson, R. W., and Dulak, N. C. (1972). *In* "Growth, Nutrition, and Metabolism of Cells in Culture" (G. H. Rothblat and V. J. Cristofalo, eds.), Vol. 1, pp. 50–81. Academic Press, New York.

Thomas, G., Martin-Pérez, J., Siegmann, M., and Otto, A. M. (1982). *Cell* **30**, 235–242.

Thompson, D. A., Gregory, L., and Ferguson-Miller, S. (1985). *J. Inorg. Biochem.* **23**, 357–364.

Todaro, G. J., and Green, H. (1963). *J. Cell Biol.* **17**, 299–313.

Trumpower, B. L. (1981). *Biochim. Biophys. Acta* **639**, 129–155.

Trumpower, B. L., and Edwards, C. A. (1979). *J. Biol. Chem.* **254**, 8697–8706.

Tse, C.-M., Belt, J. A., Jarvis, S. M., Paterson, A. R. P., Wu, J.-S., and Young, J. D. (1985). *J. Biol. Chem.* **260**, 3506–3511.

Tsuchiya, T., Misawa, A., Miyake, Y., Yamasaki, K., and Niiya, S. (1982). *FEBS Lett.* **142**, 231–234.

Tucker, R. F., Volkenant, M. E., Branum, E. L., and Moses, H. L. (1983). *Cancer Res.* **43**, 1581–1586.

Tucker, R. F., Shipley, G. D., and Moses, H. L. (1984). *Science* **226**, 705–707.

Tuy, F. P. D., Henry, J., Rosenfeld, C., and Kahn, A. (1983). *Nature (London)* **305**, 434–438.

Tzagoloff, A., ed. (1982). "Mitochondria." Plenum, New York.

Tzagoloff, A., and Penefsky, H. S. (1971). *In* "Methods in Enzymology" (W. B., Jakoby, ed.), Vol 22, pp. 219–230. Academic Press, New York.

Tzartos, S. J., and Changeaux, J.-P. (1982). *Abstr. Soc. Neurosci. 12th Annu Meet.*, p. 334.

Unden, G., and Kröger, A. (1982). *Biochim. Biophys. Acta* **682**, 258–263.

Unden, G., Mörschel, E., Bokranz, M., and Kröger, A. (1983). *Biochim. Biophys. Acta* **725**, 41–48.

Uyeda, K., and Racker, E. (1965). *J. Biol. Chem.* **240**, 4682–4693.

Vainstein, A., Hershkovitz, M., Israel, S., Rabin, S., and Loyter, A. (1984). *Biochim. Biophys. Acta* **773**, 181–188.

Van de Stadt, R. J., De Boer, B. L., and van Dam, K. (1973). *Biochim. Biophys. Acta* **292**, 338–349.

Van Heyningen, W. E. (1963). *J. Gen. Microbiol.* **31**, 275–287.

Van Hoogevest, P., van Duijn, G., Batenburg, A. M., De Kruijff, B., and De Gier, J. (1983). *Biochim. Biophys. Acta* **734**, 1–17.

Van Jagow, G., and Sebald, W. (1980). *Annu. Rev. Biochem.* **49**, 281–314.

Vara, F., and Serrano, R. (1982). *J. Biol. Chem.* **257**, 12826–12830.

Velours, J., Esparza, M., Hoppe, J., Sebald W., and Guérin, B. (1984). *EMBO J.* **3**, 207–212.

Vignais, P. V. (1976). *Biochim. Biophys. Acta* **456**, 1–38.

Villegas, R., Villegas, G. M., Barnola, F. V.,and Racker, E. (1977). *Biochem. Biophys. Res. Commun.* **79**, 210–217.

Volsky, D. J., Shapiro, I. M., and Klein, G. (1980). *Proc. Natl. Acad. Sci. U.S.A.* **77**, 5453–5457.

Waismann, D. M., Gimble, J. M., Goodman, D. B. P., and Rasmussen, H. (1981). *J. Biol. Chem.* **256**, 415–419.

Wakabayashi, S., and Goshima, K. (1982). *Biochim. Biophys. Acta* **693**, 125–133.

Walker, J. E., Runswick, M. J., and Saraste, M. (1982). *FEBS Lett.* **146**, 393–396.

Walker, J. E., Saraste, M., and Gay, N. J. (1984). *Biochim. Biophys. Acta* **768**, 164–200.

Walsh-Reitz, M. M., Toback, F. G., and Holley, R. W. (1984). *Proc. Natl. Acad. Sci. U.S.A.* **81**, 793–796.

Wang, C.-T., Saito, A., and Fleischer, S. (1979). *J. Biol. Chem.* **254**, 9209–9219.

Warburg, O. (1930). "The Metabolism of Tumors," Constable, London.

Warnock, D. G., and Rector, F. C., Jr. (1981). *In* "The Kidney" (B. M. Brenner and F. C. Rector, eds.), pp. 440–494. Saunders, Philadelphia, Pennsylvania.

Warnock, D. G., Greger, R., Dunham, P. B., Benjamin, M. A., Frizzell, R. A., Field, M., Spring, K. R., Ives, H. E., Aronson, P. S., and Seifter, J. (1984). *Fed. Proc., Fed. Am. Soc. Exp. Biol.* **43**, 2473–2487.

Watts, A. (1981). *Nature (London)* **294**, 512–513.

Wehrle, J. P., Cintrón, N. M., and Pedersen, P. L. (1978). *J. Biol. Chem.* **253**, 8598–8603.

Weigele, J. B., and Barchi, R. L. (1982). *Proc. Natl. Acad. Sci. U.S.A.* **79**, 3651–3655.

Weinberg, R. A. (1984). *Trends Biochem. Sci.* **9**, 131–133.

Weiner, J. H., Lemire, B. D., Jones, R. W., Anderson, W. F., and Scraba, D. G. (1984). *J. Cell. Biochem.* **24**, 207–216.

Weisenthal, L. M., Marsden, J. A., Dill, P. L., and Macaluso, C. K. (1983). *Cancer Res.* **43**, 749–757.

Weiss, H., Wingfield, P., and Leonard, K. (1979). *In* "Membrane Bioenergetics" (C. P. Lee, G. Schatz, and L. Ernster, eds.), pp. 119–132. Addison-Wesley, Reading, Massachusetts.

Wheeler, T. J., and Hinkle, P. C. (1982). *In* "Membranes and Transport" (A. N. Martonosi, ed.), Vol. 2, pp. 161–167. Plenum, New York.

Wheeler, T. J., and Hinkle, P. C. (1985). *Annu. Rev. Physiol.* **47**, 503–517.

Whittam, R., and Ager, M. E. (1965). *Biochem. J.* **97**, 214–227.

Wickner, W. (1976). *Proc. Natl. Acad. Sci. U.S.A.* **73**, 1159–1163.

Wickner, W. (1979). *Annu. Rev. Biochem.* **48**, 23–45.

Widdicombe, J. H., and Welsh, M. J. (1985). *Fed. Proc.* **44**, 1365, Abst. 5583.

Wikström, M. (1981). *in* "Chemiosmotic Proton Circuits in Biological Membranes (V. P. Skulachev and P. C. Hinkle, eds.), pp. 171–180. Addison-Wesley, Reading, Massachusetts.

Wikström, M., and Penttilä, T. (1982). *FEBS Lett.* **144**, 183–189.

Wikström, M., Krab, K., and Saraste, M. (1981a). *Annu. Rev. Biochem.* **50**, 623–655.

Wikström, M., Saraste, M., and Krab, K. (1981b). "Cytochrome Oxidase—A Synthesis." Academic Press, New York.

Wikström, M. K. F. (1977). *Nature (London)* **266**, 271–273.

Wikström, M., and Saari, H. (1975). *Biochim. Biophys. Acta* **408**, 170–179.

Willingham, M. C., and Pastan, I. H. (1982). *J. Cell. Biol.* **94**, 207–212.

Wilson, D. B. (1978). *Annu. Rev. Biochem.* **47**, 933–965.

Wilson, T., Papahadjopoulos, D., and Taber, R. (1979). *Cell* **17**, 77–84.

Winget, G. D., Kanner, N., and Racker, E. (1977). *Biochim. Biophys. Acta* **460**, 490–499.

Winzler, R. J. (1969). *In* "Red Cell Membrane: Structure and Function" (G. A. Jamieson and T. G. Greenwalt, eds.). p. 157. Lippincott, Philadelphia, Pennsylvania.

Wohlhueter, R. M., and Plagemann, P. G. W. (1982). *Biochim. Biophys. Acta* **689**, 249–260.

Wohlhueter, R. M., Marz, R., and Plagemann, P. G. W. (1978). *J. Membr. Biol.* **42**, 247–264.

Wohlrab, H. (1980). *J. Biol. Chem.* **255**, 8170–8173.

Wohlrab, H., and Flowers, N. (1982). *J. Biol. Chem.* **257**, 28–31.

Wong, T.-K., Nicolau, C., and Hofschneider, P. H. (1980). *Gene* **10**, 87–94.

Wright, J. K., Schwarz, H., Straub, E., Overath, P., Bieseler, B., and Beyreuther, K. (1982). *Eur. J. Biochem.* **124**, 545–552.

Wu, J.-S. R., Kwong, F. Y. P., Jarvis, S. M., and Young, J. D. (1983). *J. Biol. Chem.* **258**, 13745–13751.

Wu, R., and Racker, E. (1959). *J. Biol. Chem.* **234**, 1036–1041.

Xie, X.-S., Stone, D. K., and Racker, E. (1983). *J. Biol. Chem.* **258**, 14834–14838.

Xie, X.-S., Stone, D. K., and Racker, E. (1984). *J. Biol. Chem.* **259**, 11676–11678.

Yoshida, M., Sone, N., Hirata, H., Kagawa, Y., and Ui, N. (1979). *J. Biol. Chem.* **254**, 9525–9533.

Young, J. D. and Jarvis, S. M. (1983). *Biosci. Rep.* **3**, 309–322.

Young, J. D.-E., Blake, M., Mauro, A., and Cohn, Z. A. (1983). *Proc. Natl. Acad. Sci. U.S.A.* **80**, 3831–3835.

Younis, H. M., Winget, G. D., and Racker, E. (1977). *J. Biol. Chem.* **252**, 1814–1818.

Younis, H. M., Telford, J. N., and Koch, R. B. (1978). *Pestic. Biochem. Physiol.* **8**, 271–277.

Yu, K.-T., and Czech, M. P. (1984). *J. Biol. Chem.* **259**, 5277–5286.

Zaheer, A., Elting, J., and Montgomery, R. (1981). *J. Biol. Chem.* **256**, 1786–1792.

Zalman, L. S., Nikaido, H., and Kagawa, Y. (1980). *J. Biol. Chem.* **255**, 1771–1774.

Zilberstein, D., Agmon, V., Schuldiner, S., and Padan, E. (1982). *J. Biol. Chem.* **257**, 3687–3691.

Zimniak, P., and Barnes, E. M., Jr. (1983). *Arch. Biochem. Biophys.* **220** 247–252.

Zimniak, P., and Racker, E. (1978). *J. Biol. Chem.* **253**, 4631–4637.

Index